Dra

Rajeev Namboothiri

Drayage Operations at Seaports

A mathematical optimization framework

AV Akademikerverlag

Impressum/Imprint (nur für Deutschland/only for Germany)
Bibliografische Information der Deutschen Nationalbibliothek: Die Deutsche
Nationalbibliothek verzeichnet diese Publikation in der Deutschen Nationalbibliografie;
detaillierte bibliografische Daten sind im Internet über http://dnb.d-nb.de abrufbar.
Alle in diesem Buch genannten Marken und Produktnamen unterliegen warenzeichen-,
marken- oder patentrechtlichem Schutz bzw. sind Warenzeichen oder eingetragene
Warenzeichen der jeweiligen Inhaber. Die Wiedergabe von Marken, Produktnamen,
Gebrauchsnamen, Handelsnamen, Warenbezeichnungen u.s.w. in diesem Werk berechtigt
auch ohne besondere Kennzeichnung nicht zu der Annahme, dass solche Namen im Sinne
der Warenzeichen- und Markenschutzgesetzgebung als frei zu betrachten wären und
daher von jedermann benutzt werden dürften.

Coverbild: www.ingimage.com

Verlag: AV Akademikerverlag GmbH & Co. KG
Heinrich-Böcking-Str. 6-8, 66121 Saarbrücken, Deutschland
Telefon +49 681 9100-698, Telefax +49 681 9100-988
Email: info@akademikerverlag.de

Herstellung in Deutschland (siehe letzte Seite)
ISBN: 978-3-639-41359-5

Imprint (only for USA, GB)
Bibliographic information published by the Deutsche Nationalbibliothek: The Deutsche
Nationalbibliothek lists this publication in the Deutsche Nationalbibliografie; detailed
bibliographic data are available in the Internet at http://dnb.d-nb.de.
Any brand names and product names mentioned in this book are subject to trademark,
brand or patent protection and are trademarks or registered trademarks of their respective
holders. The use of brand names, product names, common names, trade names, product
descriptions etc. even without a particular marking in this works is in no way to be
construed to mean that such names may be regarded as unrestricted in respect of
trademark and brand protection legislation and could thus be used by anyone.

Cover image: www.ingimage.com

Publisher: AV Akademikerverlag GmbH & Co. KG
Heinrich-Böcking-Str. 6-8, 66121 Saarbrücken, Germany
Phone +49 681 9100-698, Telefax +49 681 9100-988
Email: info@akademikerverlag.de

Printed in the U.S.A.
Printed in the U.K. by (see last page)
ISBN: 978-3-639-41359-5

To my parents and brother,

living 9277 miles away, without whose unconditional love and support,

this thesis would have never been possible.

ACKNOWLEDGEMENTS

It's been a path worth taking! A journey in which I learned so much - both academically and non-academically; and a journey in which I met a lot of great minds professionally and made a lot of wonderful friends personally!!

First of all, I would like to thank my advisor Dr. Alan Erera. He is one person who has been with me throughout this journey - guiding me onto choosing a path during the initial days and providing me the independence and freedom to develop my ideas and thoughts once I was set on a path. He has been an epitome of patience - ready to listen to any problem I had as a student; and in helping me find a solution to it. He provided a very flexible and informal work environment for me, which made me feel very comfortable working with him all these years.

I would like to thank Prof. Chip White for serving in my committee and for his valuable guidance during my initial days as a member of the SETRA research group, and for helping me in shaping up a lot of my research ideas while working in that group. I would like to thank Prof. Martin Savelsbergh for serving in my committee, and for his inputs and suggestions during our meetings, which definitely provided more avenues to think about and explore. It was also great pleasure working with him as a part of ISyE 7653. I would like to thank Prof. John Vande Vate for serving in my committee, and for providing interesting ideas to explore since the thesis proposal stage. I would also like to thank Prof. Soumen Ghosh for serving in my committee, and in being greatly flexible. It was a wonderful experience, learning 'Global Operations' from him through MGT 6360.

I would like to thank the Sloan Foundation's Trucking Industry Program in providing me the financial support to conduct this research.

Finally, I would like to thank all my roommates and friends for creating a homely atmosphere, so far away from home. The list of names is too long to include here. With them being around, the personal life aspect of life at grad school, which has often been criticized for various reasons, was a surprisingly refreshing experience for me. These people made my entire stay at grad school a memorable experience - all those friday night movies, the restaurant/movie outings etc will be fondly remembered.

TABLE OF CONTENTS

LIST OF TABLES

LIST OF FIGURES

SUMMARY

This dissertation considers daily operations management for a fleet of trucks providing container pickup and delivery service to a port. Truck congestion at access points for ports may lead to serious inefficiencies in drayage operations, and the resultant cost impact to the intermodal supply chain can be significant. Recognizing that port congestion is likely to continue to be a major problem for drayage operations given the growing volume of international containerized trade, this research seeks to develop optimization approaches for maximizing the productivity of drayage firms operating at congested seaports. Specifically, this dissertation addresses two daily drayage routing and scheduling problems.

In the first half of this dissertation, we study the problem of managing a fleet of trucks providing container pickup and delivery service to a port facility that experiences different access wait times depending on the time of day. For this research, we assume that the wait time can be estimated by a deterministic function. We develop a time-constrained routing and scheduling model for the problem that incorporates the time-dependent congestion delay function. The model objective is to find routes and schedules for drayage vehicles with minimum total travel time, including the waiting time at the entry to the port due to congestion. We consider both exact and heuristic solution approaches for this difficult optimization problem. Finally, we use the framework to develop an understanding of the potential impact of congestion delays on drayage operations, and the value of planning with accurate delay information.

In the second half of this dissertation, we study methods for managing a drayage fleet serving a port with an appointment-based access control system. Responding to growing access congestion and its resultant impacts, many U.S. port terminals have implemented appointment systems, but little is known about the impact of such systems on drayage productivity. To address this knowledge gap, we develop a drayage operations optimization approach based on a column generation integer programming heuristic that explicitly models a time-slot port access control system. The approach determines pickup and delivery sequences with minimum transportation cost. We use the framework to develop an understanding of the potential efficiency impacts of access appointment systems on drayage operations. Findings indicate that the set of feasible drayage tasks and the fleet size required to complete them can be quite sensitive to small changes in time-slot access capacities at the port.

CHAPTER I

INTRODUCTION

1.1 Background

Continuous growth in global trade volumes has strained the transportation infrastructure supporting international trade. In the United States, substantial container volume growth has made it more difficult for ports to efficiently serve drayage trucks entering and exiting with containers, especially during peak periods. Queues of drayage trucks frequently form both at port entrance gates and also within the facility at container pickup and dropoff points. While on-facility rail connections help alleviate truck congestion at some ports, demand for trucking service is likely to remain at high levels.

Growth in U.S. container trade is not expected to slow. In 2001, a top government official echoed the prevailing views of economists in testimony before a joint hearing of the House Transportation and Infrastructure Committee when he speculated that total U.S. international trade tonnage will double by the year 2010, of which 95% will move through U.S. ports (Herberger, 2001). In this testimony, the official refers to the prediction provided by the firm of DRI McGraw-Hill that U.S. international container trade will nearly double during this time frame. Given that about 16 million container TEUs (twenty foot equivalent units) enter and exit the U.S. annually, this growth rate of course implies that an additional 16 million additional TEUs will be handled by U.S. ports in the near future. Managing this growth in traffic, and the additional congestion it brings, will clearly be an important task for all port stakeholders.

New security mandates may also compound port congestion problems. One important example mandate is the new Transportation Worker Identification Credentialing (TWIC) program, which requires all U.S. ports to verify the identity of each individual accessing their facilities, including drayage truck drivers. These credential systems are designed to verify the identity of workers, validate their background information, and create an audit trail for port access. In 2002 testimony before the House Transportation and Infrastructure Committee, a senior official in a maritime trade organization pointed out that performing manual personal identification or cargo security checks will slow down the flow of cargo through the ports with little resultant increase in security (MacDonald, 2002). Until an efficient credentialing system is adopted by ports, verifying proper credentials at port gates is likely to contribute to access delays. Top officials recognize that providing security and efficiency simultaneously should be a high priority goal in the design of improvements to seaports (Schubert, 2002).

1.2 Port Operations

Figure 1 provides an overview of the various activities performed at a port by drayage trucks. Trucks arrive at the port entry gate, either with a container, with an empty chassis, or as a bobtail. If there is a queue of trucks at the entry gate, some waiting occurs. Once processed and on the port grounds, the truck proceeds to a dropoff location if arriving with a container, or to a pickup location. Trucks arriving as a bobtails must first proceed to a chassis pickup location, where drivers inspect available chassis and make a selection. Trucks may wait in queue for yard cranes to load outbound containers at pickup locations. Finally, all trucks leaving the facility must again be processed through the access gates. Each of these activities requires processing time, and potentially some waiting time when the port is congested. During peak hours of the day, and during peak times of the year, these congestion delays can be quite significant.

Port operators, container shippers, and drayage trucking companies all make operational decisions that affect the ability of a seaport to efficiently manage container throughput. For example, operators decide on truck access gate operating hours and staffing levels, and also allocate and manage resources devoted to various container and chassis handling functions within the port. Container shippers specify when they would like containers delivered to and picked up from the port, usually by setting pickup and delivery time windows or deadlines. Finally, drayage trucking companies make routing and scheduling decisions that in turn determine the precise times when containers are picked up from or delivered to the seaport.

2

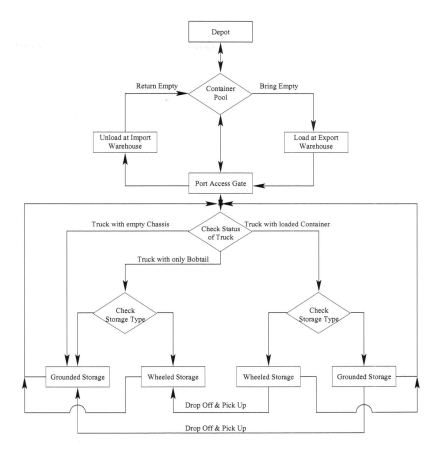

Figure 1: Activities performed by drayage vehicles serving a port

Some research studies have been recently conducted in the attempt to understand the impact of port congestion on the efficiency of trucking firms providing drayage access to maritime intermodal facilities. Regan and Golob (2000) presents the results of a survey of nearly 1200 private and for-hire carriers operating in California. Over 450 of the companies surveyed had operations involving maritime ports in California. More than 40% of the surveyed dispatch and operations managers claim that drivers typically wait more than an hour outside the port prior to access. Additionally, 75% claim that drivers typically spend more than an hour inside the port. Waiting times were also typically quite variable; over 80% said that the time the driver would spend at the port was not predictable to within 30 min. This combination of delays and variability led 44% of the firms to report that operations were often significantly affected by congestion. Importantly, Regan and Golob (2000) postulates that customer time windows for pickup and delivery compel drayage operators to work during congested time periods when it would be better to wait for less busy periods. Such delays clearly reduce the efficiency of the drayage company, since more driver resources may be required to complete a set of container movement tasks.

Most, if not all, ports in the U.S. operate unscheduled drayage access; essentially, trucks may arrive to pickup and drop off containers any time within gate operating hours. This policy may be inefficient, since there may be time periods during the day when resources are idle and others when resources are oversaturated. Excessive queuing caused by this policy has also begun to worry government officials in areas with major seaports, mostly motivated by the environmental hazard created by diesel engine emissions from idling trucks waiting outside port gates. One proposed remedy for the environmental and productivity problems generated by unscheduled access are truck access appointment systems. Recently, the state legislature in California implemented the so-called Lowenthal bill (Assembly Bill 2650), which proposed restricting the allowable time trucks can idle in port terminals to 30 minutes. This bill also proposed the implementation of an appointment system, which would ensure that trucks not idle or queue for more than 30 minutes while waiting to enter the gate into the marine terminal.

In response, many West coast terminals have now implemented an appointment system which allows truckers to schedule arrival appointments at the gate. Recent trade news articles have reported reduced wait time for motor carriers, reduced operational costs for terminals from improved gate efficiency, accelerated throughput, and better equipment utilization.The Port of Singapore Authority (PSA) also operates an appointment system for drayage access to its terminals. According to Rajasimhan (2002), at PSA Singapore Terminals the flow-through gate system processes one truck every 25 seconds. On any given

4

day, about 8,000 container trucks pass through the gates of PSA's four terminals. During peak periods, over 700 trucks are processed each hour.

For trucking companies accessing terminals with appointment systems, these benefits hopefully lead to increased "turns" and enhanced profitability. However, given the additional constraints such systems place on operations, careful planning is required to attain the maximum benefit.

Given the expected surge in international container trade coupled with its resultant strain on the port infrastructure, it is timely to investigate whether existing infrastructure might be utilized more efficiently. Operational decision support for drayage trucking firms may be able to relieve some of the burden of growing volumes by enabling efficiency gains. Some research has previously explored using information technology solutions to improve drayage operations at ports. Holguin-Veras and Walton (1996) and Holguin-Veras (2000) study the feasibility of the use of information technology in port operations by interviewing port operators and by conducting a small survey of carriers. These studies conclude that although port and terminal operators frequently use information technologies to enhance internal efficiency, trucking firms providing drayage access were reluctant to follow suit. Container status information systems were perceived to be too costly and unreliable. It is not clear, however, that drayage carriers understand the benefits of systems supported by information technology; the study in this dissertation serves partially to address this concern.

1.3   Research Focus

This dissertation will focus on improving decision-making by drayage trucking firms serving congested seaports, and understanding the productivity impacts of seaport congestion and congestion control measures. Providing service to congested facilities presents a challenge for drayage firms, since their productivity can be significantly compromised by the driver idle time generated by delays. In the research that follows, we investigate routing and scheduling decision-making for a drayage firm under two scenarios. In the first, we assume that the drayage firm has a reasonable estimate of the expected congestion delay that it will encounter over the course of a day. Given these expected delays, the goal of the firm is to develop container routes and schedules that maximize productivity. In the second scenario, we consider a drayage firm providing service to a port that has implemented an appointment-based access control system to manage congestion. In such a system, the port operator limits the number of truck moves allowed by the firm into the port facility at different times of the day. Given such a system, again the goal of the drayage firm is to develop routes and schedules that maximize productivity without violating the constraints of the access control system.

In the remainder of this introduction, we first define the specific container drayage problem setting that we will use as a basis for the research. Next, we outline the specific contributions of the dissertation research. Finally, we provide a roadmap to the remaining chapters of the dissertation.

1.4   Problem Setting

Consider the operational planning problem faced by a trucking company serving local drayage container moves to and from a single intermodal port. Suppose that the company operates a fleet of tractors and drivers based at a single nearby depot. The company receives container move requests from export and import customers. Given a set of container move requests, the company would like to determine a minimum-cost set of vehicle schedules serving all orders during a planning period (*e.g.*, one day). Cost is assumed to be a linear function of the number of vehicles required to serve the requests and the total travel distance required by all vehicles. Further, assume that the company has advance knowledge of the complete set of container requests for the period. Although there may be real-world scenarios where some orders arrive dynamically during the planning period, this consideration is ignored in this research. We also assume that any driver-tractor combination can serve any move request.

As mentioned above, container move requests are generated by exporters and importers, or parties organizing logistics on their behalf (*e.g.*, ocean carriers or third-party logistics providers). For exposition, in this dissertation we define exporters as customers that send a container to the port, and importers as those that receive a container from a port. Move requests may represent empty or loaded containers, however we assume that all empty container moves must begin or end at the port. It is often the case that empty containers are hauled in conjunction with loaded moves; for example, an exporter might receive an empty container and then send the truck back to the port with a load creating a round-trip. Such a case is modeled with two container move requests: an empty import request, and then a loaded export request. Note that we do not explicitly model precedence relationships between such linked tasks; we assume that such precedence is captured implicitly in the request time window information.

Each container request includes an origin location where the container is to be picked up and a destination location where the container is to be dropped off; by definition, the port location is either the origin or destination of each request. Each request may also specify time window information at both the origin and destination. The time window at the origin specifies the earliest and latest pickup times and at the destination the earliest and latest delivery times for the container. In this research, we treat these time windows as hard constraints, violation of which results in an infeasible route. Each vehicle moves a single container at a time in drayage operations, and we assume for this local problem that each vehicle begins and ends its route each day at a single fixed depot location. Further, we assume that travel times between locations are known with certainty. For some of problems to be studied, vehicles will be occupied for some time at certain locations, and these occupation times (or service times) will be known with certainty but may be vary in duration over the planning period.

Figure 2 depicts a typical network associated with port drayage, and two example vehicle drayage routes. In this example, node P is the port location, D is the depot location, E_1, E_2, E_3 and E_4 are the export customer locations and I_1, I_2 and I_3 are the import customer locations. As observed earlier, each container move, denoted by solid lines, begins or terminates at the port location P. The two different dotted lines depict routes performed by two different vehicles, and these represent typical routes associated with drayage trucks in practice. The route for vehicle 1 is given by $D \rightarrow E_4 \rightarrow P \rightarrow D$ and the route for vehicle

2 is $D \to P \to I_1 \to E_1 \to P \to I_3 \to D$.

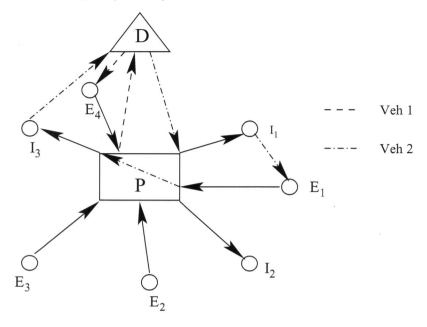

Figure 2: Example local drayage network

1.5 Contributions

The primary contributions of this dissertation are:

- Development and thorough analysis of a fast, polynomial-time heuristic denoted the Layered Shortest Path (LSP) heuristic that generates near-optimal solutions to the elementary shortest path problem with resource constraints (ESPPRC); the class of column generation subproblem encountered in this research.

- Development of a mixed-integer programming formulation for a basic drayage routing and scheduling problem; and for drayage routing and scheduling problems with known piecewise-linear time-dependent waiting time for port access.

- Construction of proofs that uncongested drayage routing and scheduling problems, which are special cases of several well-known hard problems, are in the class NP-hard; it follows that the congested generalization of the problem and the access-controlled generalization are also NP-hard. These results motivate the search for effective solution procedures;

- Development of decision support models based on column generation, by extending the fast heuristics developed for the generic ESPPRC column generation subproblem, for a basic drayage routing and scheduling problem; for drayage routing and scheduling problems that incorporate known time-dependent waiting time for port access; and for port drayage routing and scheduling problems with port access time slot capacities.

- Computational studies investigating the quality of solutions and the computational efficiency of using a standard root node column generation heuristic for solving the VRPTW, with the LSP heuristic used to solve the pricing subproblem. Results indicate that the LSP approach is a computationally attractive method for problems of practical size that yields good to very good solution quality, especially when the number of customers per tour is small.

- Computational studies investigating the impact of port access delays on drayage operations; and on the productivity impacts of implementing an access control system.

- Computational studies investigating the importance of incorporating expected delays into planning decisions; and on the importance of designing optimal parameters for an access control system.

1.6 Roadmap

The rest of this dissertation is organized as follows. Chapter 2 introduces a fast heuristic approach for solving Elementary Shortest Path Problems with Resource Constraints (ESP-PRC), which will be utilized in all of the heuristics developed in this dissertation. First, we provide a brief overview of existing techniques for solving ESPPRC. We then develop an alternative approach, denoted the Layered Shortest Path (LSP) heuristic, designed to

9

work efficiently to handle problems with less restrictive resource constraints and therefore many more feasible solutions. Finally, we test the LSP heuristic on Solomon's benchmark instances for VRPTW, and show that the quality of solution obtained is comparable to other standard techniques.

Chapter 3 discusses an uncongested container drayage problem that will serve as the base decision problem for the remaining research. The problem definition is followed by a review of existing literature for similar problems. Complexity results are provided. The problem is then formulated as a mixed-integer linear program. Solution procedures based on a set partitioning approach with various column generation heuristics are developed and presented. Computational results demonstrate the effectiveness of the methods developed.

Chapter 4 introduces a generalization of the uncongested drayage problem that allows the modeling of time-dependent deterministic port access delay. A survey of existing literature on routing problems with time-dependent travel times is provided. Extensions of the solution procedures developed in Chapter 3 for the congested drayage problem are discussed, supported by computational results. This chapter also presents a computational study to show the impact of congestion on the drayage operations of a single firm. The results emphasize the value of congestion delay information when planning drayage operations.

Finally, Chapter 5 introduces an alternative generalization of the uncongested drayage problem in which access to the port is limited by a time-slot based port capacity appointment system. Appointment systems in place at many ports worldwide are first discussed. Then, we formally define the mathematical optimization problem that we will consider for the remainder of this chapter. Finally, we discuss solution approaches for the problem based on modifications of the methods developed for generalized routing problems with resource constraints presented in Chapter 2. Computational results demonstrate the sensitivity of drayage operational productivity to available access capacity.

CHAPTER II

A LAYERED SHORTEST PATH HEURISTIC FOR ROUTING COLUMN GENERATION

2.1 Introduction

In this chapter, we introduce a heuristic solution technique for the Elementary Shortest Path Problem with Resource Constraints (ESPPRC). Instances of this problem are frequently encountered as subproblems in column generation approaches for solving routing problems with resource constraints, such as the Vehicle Routing Problem with Time Windows (VRPTW). The solution heuristic that we develop is denoted the Layered Shortest Path (LSP) heuristic. Alternatively to other approaches for finding optimal or suboptimal solutions to the ESPPRC, the LSP heuristic is a fast polynomial approach for finding elementary paths that are feasible with respect to the resource constraints; the most common approach in research practice utilizes a pseudo-polynomial dynamic programming technique to find (not necessarily elementary) paths.

The remainder of this chapter is organized as follows. First, we give a brief overview of various approaches developed in the research literature for developing both exact and suboptimal solutions to the ESPPRC. Next, we describe and analyze the LSP heuristic, and compare it to existing solution approaches. Finally, we test a root column generation heuristic for the VRPTW using the LSP heuristic to price columns on Solomon's benchmark instances, and show that the method produces solutions of high quality with good computational efficiency; the performance is comparable to other standard techniques.

2.2 Background Literature

In many routing and scheduling problems with resource constraints solved by column generation, the pricing subproblem for identifying solution-improving columns to the linear relaxation of a set partitioning or covering integer program corresponds to a shortest path problem with resource constraints (SPPRC) or one of its variants. Irnich and Desaulniers (2005) provides a comprehensive survey on the subject, and proposes a classification and a generic formulation for SPPRCs. The reference also briefly discusses complex modeling issues involving resources, and presents the most commonly employed SPPRC solution methods.

The classic optimization approach for the VRPTW is given in Desrochers et al. (1992). The paper formulates the problem using a set partitioning model, and proposes to solve its linear relaxation using column generation. Although the pricing subproblem is a variant of the SPPRC known as the Elementary SPPRC (ESPPRC), the paper relaxes the elementary path condition and develops a pseudo-polynomial optimal algorithm based on dynamic programming for its solution. The pseudo-polynomial complexity is likely to have the best theoretical worst-case computational requirements. The complete approach proposed is able to generate optimal solutions to some of the 100-customer Solomon VRPTW instances.

Dror (1994) proves that the ESPPRC subproblem for the set partitioning formulation of the VRPTW is NP-hard in the strong sense.

As an alternative to the classic approach of Desrochers et al. (1992), D. Feillet and Gueguen (2004) proposes an exact solution algorithm for the ESPPRC. This algorithm is based on the label correcting algorithm proposed in Desrochers (1988), with new resources in the label structure to enforce the elementary path constraint. Again, in order to solve the problem exactly, the paper resorts to a pseudo-polynomial approach which maintains resource consumption information in the state space.

Recently, some research has focused on using constraint programming to solve the pricing subproblems in the routing context. For example, Rousseau et al. (2004) presents a column generation approach that solves the elementary shortest path subproblem with constraint programming. The proposed method is flexible since it can handle not only resource based constraints but almost any structure of constraints, while still providing acceptable performance on known benchmark problems.

2.3 Column Generation Approach for Routing with Resource Constraints

When routing problems include binding resource constraints that limit the number of feasible routes, it is often appropriate to utilize set covering integer programming formulations for their solution. We are going to use a set covering type model, as the linear relaxation of the set covering type model is numerically far more stable than that of a set partitioning model. The set covering approach is a type of partial enumeration, in which single-vehicle routes are enumerated, but then integer programming is used to select a subset of the set of routes that covers all customer demands and optimizes some objective. Complete enumeration of *all* feasible single-vehicle routes is usually impractical for the large routing problem instances encountered in practice. Hence, for large instances, a column generation approach is generally employed. In this approach, column generation is used to solve the linear programming relaxation of a set covering routing formulation; column generation is either employed only at the root node (in heuristic approaches), or at all nodes in the branch-and-bound tree (in exact approaches). The general approach is detailed in this section.

2.3.1 Set Covering Routing Formulation

Resource-constrained routing and scheduling problems can be formulated using set covering formulations when the operations of individual vehicles can be treated separately except for linking constraints that guarantee that each customer is served by exactly one vehicle. Given the set $\mathcal{C}$ of all customer requests ($n = |\mathcal{C}|$) and the set $\mathcal{R}$ of all feasible single-vehicle routes serving subsets of $\mathcal{C}$, a set covering model can be solved to determine the minimum-cost subset of $\mathcal{R}$ which ensures that each customer request is served.

Consider a routing problem where the objective is to minimize the total cost required by all operated routes. Let α_{ij} be a $\{0, 1\}$ parameter equal to one if request i is served by route j, and let c_j be the total cost required by route j. The decision variables x_j indicate which routes in $\mathcal{R}$ are chosen for the optimal subset:

$$
x_j = \begin{cases} 1 & \text{if route } j \text{ is in final optimal solution} \\ 0 & \text{otherwise} \end{cases}
$$

The set covering binary integer programming formulation $P(\mathcal{R})$ is then:

$$
\text{minimize} \quad \sum_{j \in \mathcal{R}} c_j x_j
$$

subject to:

$$
\sum_{j \in \mathcal{R}} a_{ij} x_j \geq 1 \quad \forall \ i \in \mathcal{C} \tag{1}
$$

$$
x_j \in \{0, 1\} \quad \forall \ j \in \mathcal{R} \tag{2}
$$

Since $\mathcal{R}$ will generally contain a very large number of routes for instances of practical size, an iterative procedure based on column generation, initially described in Dantzig and Wolfe (1960), is employed to solve problem $P(\mathcal{R})$.

2.3.2 Column Generation Solution Approaches

In column generation solution approaches to $P(\mathcal{R})$, the linear programming relaxation of the set covering model is solved initially with a subset of columns $\mathcal{R}'$, each $r \in \mathcal{R}'$ representing a feasible single-vehicle route; denote this problem $P^{LP}(\mathcal{R}')$. Importantly, the subset $\mathcal{R}'$ should be one such that there exists a feasible solution to $P^{LP}(\mathcal{R}')$. Then, using optimal dual variable information from the solution of this restricted linear relaxation, a pricing subproblem is solved to either generate columns with negative reduced cost to be introduced to improve the solution to the restricted linear program, or prove that no such columns exist. If found, negative cost columns are added to the restricted problem and it is solved again; this process is continued until the subproblem ceases to find any additional columns with negative reduced cost, at which point the optimal solution to the linear relaxation has been determined.

Using column generation to solve integer programming problems to optimality requires that an optimal pricing algorithm is applied to generate columns, and that the column generation is repeated for each of the linear programming problems solved throughout the branch-and-bound search tree. Branch-and-price techniques are methods for extending column generation techniques throughout a branch-and-bound tree. A heuristic alternative is to only apply column generation for the root node linear relaxation; we will denote this approach a *root column generation heuristic*. Once the set of columns $\mathcal{R}'$ has been determined that yields the optimal solution at the root, the binary integer programming problem is solved using only columns $\mathcal{R}'$. Such an approach should work reasonably well, especially for problems with large numbers of customers. In the case of VRPTW, Bramel and Simchi-Levi (1997) studies the gap between the linear programming relaxation solution and the optimal integer solution of the set covering model. The paper demonstrates that for any distribution of service times, time windows, customer loads, and locations, the relative gap between fractional and integer solutions of the set covering problem becomes arbitrarily small as the number of customers increases.

A generic root column generation heuristic to solve a set covering integer programming formulation of a routing problem is presented below:

Root Column Generation Heuristic

1: Let $\mathcal{R}'$ be the set of all routes covering exactly one customer request

2: **repeat**

3:    Solve $P^{LP}(\mathcal{R}')$, the linear relaxation of set covering model using route subset $\mathcal{R}'$

4:    Solve pricing subproblem and add to $\mathcal{R}'$ all routes with negative reduced cost $\bar{c}_j < 0$

5: **until** Pricing subproblem identifies no routes to add

6: Solve $P(\mathcal{R}')$, the binary integer set covering model using route subset $\mathcal{R}'$

As described earlier, the most frequent pricing subproblem encountered for routing set covering models is the Elementary Shortest Path Problem with Resource Constraints (ESPPRC). In the sections to follow, we present a generic formulation and a new heuristic solution method for the ESPPRC.

2.4 Elementary Shortest Path Problem with Resource Constraints

We first present a formal mathematical definition of the ESPPRC.

2.4.1 ESPPRC Definition

Consider a directed graph $G = (\mathcal{V}, \mathcal{A})$ where $\mathcal{V}$ is the set of vertices and $\mathcal{A}$ is the set of arcs. Suppose $v_0 \in \mathcal{V}$ is the start node, from which we would like to find paths. Let $\mathcal{S}$ represent a set of resources. Each vertex $v_i \in \mathcal{V}$ has a constraint associated with each resource $s \in \mathcal{S}$ which bounds the cumulative consumption of the resource along the path from v_0 to the interval $[a_{v_i}^s, b_{v_i}^s]$. When a path reaches node v_i, if the consumption of resource s is less than $a_{v_i}^s$, it is set to $a_{v_i}^s$. Note that it is straightforward to use such a construct to model time window constraints for customers. Each arc $(v_i, v_j) \in \mathcal{A}$ has a corresponding cost $\bar{c}_{ij}$, which we note may be negative. Using an arc also consumes some amount of each resource; let t_{ij}^s be the consumption of resource s on arc (v_i, v_j). It is simple to see how a time resource in a routing network can be modeled using this framework. Note also, however, that a capacity resource could also be modeled with this framework. Suppose each vertex has a demand of q_j units. Space capacity can be thus modeled using a resource interval of $[0, Q]$ at each vertex, where Q is the total capacity of the vehicle in units, and a consumption of $t_{ij}^s = q_j$ for each arc inbound to vertex j.

We now consider the problem of determining elementary paths from vertex v_0:

Definition 1 (Elementary Path). *An elementary path in a directed graph is a path in which there are no repeated vertices.*

The ESPPRC is to find a minimum cost feasible elementary path from vertex v_0 to some other vertex v_j in G, where a feasible path satisfies each of the resource constraints at every node along the path. Note that if a feasible path exists from v_0 to v_j, there must exist a minimum cost feasible elementary path since cycling is prevented by the elementary condition. Further, note that approaches for solving ESPPRC problems often find minimum cost feasible elementary paths from v_0 to a set of other vertices in G with no increase in computational complexity.

2.4.2 ESPPRC Subproblem in Column Generation

The ESPPRC problem often arises as the pricing subproblem in set cover column generation for routing problems. Consider again the set covering formulation $P(\mathcal{R})$ for a routing problem, and suppose that we are solving the linear relaxation at the root of the branch-and-bound tree. Further, suppose that we have the solution the $P^{LP}(\mathcal{R}')$, where $\mathcal{R}' \subset \mathcal{R}$. Let $\pi = \{\pi_i\}$ represent the dual variables associated with constraints (1). Then, the reduced cost $\bar{c}_j$ of any route $j \in \mathcal{R}$ is given by

$$\bar{c}_j = c_j - \sum_{i \in C} \alpha_{ij} \pi_i. \tag{3}$$

Given a set of dual variables, determining one (or many) routes in $\mathcal{R}$ with negative reduced cost is an instance of ESPPRC. To see that this is true, let $\mathcal{V} = \mathcal{C} \cup \{0\}$, where vertex 0 represents the depot. Further, let $v_0 = 0$. Modify the cost of each arc (v_i, v_j) such that $\bar{c}_{ij} = c_{ij} - \pi_j$. Establish appropriate resource consumptions and constraints, depending on the type of routing problem to be solved. Then, solve the ESPPRC to determine minimum cost feasible elementary paths from the depot v_0 to all vertices in $\mathcal{C}$. For each vertex $v_k \in \mathcal{C}$, add the cost c_{k0} to the cost of the minimum cost path to v_k to generate a reduced cost $\bar{c}_k$. If the reduced cost for any vertex v_k is negative, add to $\mathcal{R}'$ the ordered route covering all of the vertices in the path from the depot to v_k. If no reduced costs are negative, the column generation can be stopped and an optimal solution has been determined.

In the remainder of this chapter, we develop heuristic techniques for solving the ESPPRC pricing subproblem. Then, we use the approach within a root column generation procedure for standard instances of the VRPTW to demonstrate its effectiveness.

2.5   *Layered Shortest Path (LSP) Heuristic*

The Layered Shortest Path method, which we denote LSP, heuristically solves the Elementary Shortest Path Problem with Resource Constraints, described in Section 2.4. The method conducts a heuristic search for resource-constraint-feasible elementary shortest paths in the network $G = (\mathcal{V}, \mathcal{A})$. The heuristic utilizes a very fast labelling procedure over a *layered* vertex array structure Λ, depicted in Figure 3.

We denote each row i of the array Λ in Figure 3 as a *layer*. The number of layers in Λ is no greater than $|\mathcal{V}| - 1 = n$, the number of vertices other than the start vertex v_0. The set of array elements in layer k represents the set of vertices $\mathcal{V} \setminus \{v_0\}$.

17

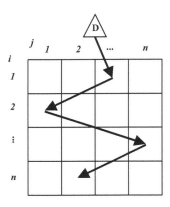

Figure 3: Layered vertex array for LSP heuristic.

Definition 2 (Layered Path). *A path P originating at vertex v_0 and destined for vertex j at layer k is a layered path, denoted $P_{v_0 j}(k)$, if it visits k vertices, exactly one from each layer of Λ.*

The LSP heuristic generates *feasible* and *elementary* layered shortest paths by extending paths quickly, layer by layer, beginning with layer $k = 1$. Let the cost of a path P be given by $c(P)$.

Definition 3 (Layered Shortest Path). *A path P originating at vertex v_0 and destined for vertex j at layer k is called a layered shortest path at layer $k > 1$, denoted $P_{v_0 j}^*(k)$, if*

$$c(P_{v_0 j}^*(k)) \leq c(P_{v_0 i}^*(k - 1)) + \bar{c}_{ij}$$

for all $i \in \mathcal{V} \setminus \{v_0\}$ such that the path $P_{v_0 i}^*(k-1)$ can be feasibly extended to j while satisfying all resource constraints. Further, the layered shortest path at layer $k = 1$, $P_{v_0 j}^*(1)$ is simply $\{v_0, j\}$ with cost $\bar{c}_{v_0 j}$, as long as the path is resource feasible.

Importantly, note that the set of resource-feasible layered shortest paths does not necessarily contain the optimal resource-feasible shortest path from v_0 to each node j. A counterexample will be provided later in the chapter.

The LSP heuristic is now described in detail.

18

The LSP heuristic generates elementary layered shortest paths by employing a label-setting approach. Each array element $(k, j) \in \Lambda$ will be used to store information about path $P^*_{v_0 j}(k)$ in a *label* ℓ_{kj}. One additional label ℓ_0 is used for initialization. Let $\mathcal{L}$ denote the set of all labels ℓ_{kj} generated by the heuristic. Each label $\ell_{kj} \in \mathcal{L}$ consists of the following attributes:

- A path vector $\mathbf{p}_{kj}$ of length n with each element

$$
p^i_{kj} = \begin{cases} 1 & \text{if vertex } v_i \in \mathcal{V} \setminus \{v_0\} \text{ is already covered in path } P^*_{v_0 j}(k) \\ 0 & \text{otherwise} \end{cases}
$$

- The current consumption of each resource s by path $P^*_{v_0 j}(k)$, given by T^s_{kj}, for $s \in \mathcal{S}$

- The cost of path $P^*_{v_0 j}(k)$, given by $\delta_{kj} = c(P^*_{v_0 j}(k))$

LSP Heuristic

Initialization:

1: Initialize label ℓ_{0,v_0} corresponding to the start node v_0 representing an empty path, set all attributes of ℓ_{0,v_0} to 0

Iterations:

2: **for all** $j \in \mathcal{V} \setminus \{v_0\}$ **do**

3:    **if** $checkExtend(\ell_{0,v_0}, j) = $ TRUE **then**

4:       $layerExtend(\ell_{0,v_0}, j)$

5:    **end if**

6: **end for**

7: $k = 2$

8: **while** $k \leq |\mathcal{V}| - 1$ **do**

9:    **for all** $i \in \mathcal{V} \setminus \{v_0\}$ such that $\ell_{k-1,i}$ exists **do**

10:       **for all** $j \in \mathcal{V} \setminus \{v_0\}$ **do**

11:          **if** $checkExtend(\ell_{k-1,i}, j) = $ TRUE **then**

12:            $layerExtend(\ell_{k-1,i}, j)$

13:        **end if**

14:      **end for**

15:    **end for**

16:    $k = k + 1$

17: **end while**

$checkExtend(\ell_{ki}, j)$

1: **if** $p_{ki}^j = 0$ **then**

2:    **if** $T_{ki}^s + t_{ij}^s \leq b_j^s \qquad \forall s \in \mathcal{S}$ **then**

3:      **if** $\ell_{k+1,j}$ does not exist **then**

4:        Generate blank label $\ell_{k+1,j}$

5:        return TRUE

6:      **else**

7:        **if** $\delta_{ki} + \bar{c}_{ij} \leq \delta_{k+1,j}$ **then**

8:          return TRUE

9:        **end if**

10:      **end if**

11:    **end if**

12: **end if**

13: return FALSE

$layerExtend(\ell_{ki}, j)$

1: $\mathbf{p}_{k+1,j} = \mathbf{p}_{ki}$

2: $p_{k+1,j}^j = 1$

3: $\delta_{k+1,j} = \delta_{ki} + \bar{c}_{ij}$

4: $T_{k+1,j}^s = max(T_{ki}^s + t_{ij}^s, a_j^s) \qquad \forall s \in \mathcal{S}$

The subroutine *checkExtend* is used to determine whether or not a path can be feasibly extended from the layered shortest path currently terminating at vertex i in layer k to a layered shortest path currently terminating at vertex j in layer $k-1$. The extension is feasible if node j is not already on the path to node i, all resource constraints are satisfied, and the extension will result in a lower cost layered path to j at layer $k+1$. If the conditions are met, then the subroutine *layerExtend* is used to update the path to j at layer $k+1$.

2.6   Analysis of LSP Heuristic

The LSP heuristic extends partial paths by the principle of reaching, *i.e.*, each optimal partial path $P^*_{v_0 i}(k)$ is considered for extension to vertices at layer $k+1$, maintaining feasibility and elementary constraints. Since layered shortest paths are determined optimally for layer k before considering any paths to layer $k+1$, permanent labels can be set for layer $k+1$ by considering at most $|\mathcal{V}| - 1$ potential extensions from layer k; thus the method is very efficient.

2.6.1   Non-optimality of LSP Heuristic

The LSP heuristic does not necessarily find all optimal shortest paths from v_0 for a given ESPPRC instance. Some potential candidate paths might be pruned at an earlier level before they can be extended to generate an optimal resource-feasible elementary shortest path. This is because the LSP heuristic stores at most one label per vertex per layer. The example below with a single source node and 3 vertices illustrates this drawback.

Figure 4 shows an example network, with the travel costs represented along each arc.

The result of running the LSP heuristic on this instance is provided in Table 1. Each entry in the table is a generated label, and labels in italics are dominated and removed at some step of the procedure. The first component of the label is the cost of the route, and the second component (in parentheses) is the previous vertex number from the prior layer.

Table 1: Optimal Label Array for LSP Heuristic Non-optimality Example

	1	2	3
1	2	2	2
2	3(2) *7(3)*	*1(1)* 0(3)	*6(1)* 0(2)
3	1(2) *5(3)*		7(1)

21

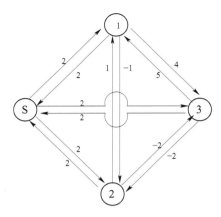

Figure 4: Network yielding non-optimal solutions to the ESPPRC when using the LSP heuristic

At the completion of the heuristic, no paths with negative cost are identified. However, route s-1-2-3 has an associated travel cost of -1. At level 2, however, the label corresponding to s-1-2 is dominated by route s-3-2. Since route s-1-2 is not extended, the method does not generate the negative cost path s-1-2-3.

2.6.2 LSP as an Unconstrained Shortest Path Algorithm

Let $G = (\mathcal{V}, \mathcal{A})$ be a directed graph with source v_0 and cost function $c : \mathcal{A} \rightarrow R$. Assume that G contains no negative cycles that are reachable from v_0. Let $d[v_j]$ represent the cost corresponding to the current shortest path from v_0 to v_j and $\eta[v_j]$ the predecessor of v_j in that path. Also, let $\delta(v_0, v_k)$ represent the cost corresponding to the optimal shortest path.

This setting represents the unconstrained version of the Shortest Path Problem with Resource Constraints. We claim that the LSP algorithm can be used to find the optimal shortest path from v_0 to each vertex $v_j \in \mathcal{V}$.

In order to prove the optimality of LSP algorithm to find shortest paths, we recall the process of relaxation and the path-relaxation property for shortest paths.

Definition 4 (Edge Relaxation). *The process of relaxing an edge (v_i, v_j) consists of testing*

whether we can improve the shortest path to v_j found so far with a path through v_i and, if so, updating $d[v_j]$ and predecessor $\eta[v_j]$.

Property 1 (Path Relaxation)**.** *If $p = (v_0, v_1, \ldots, v_k)$ is a shortest path from v_0 to v_k, and the arcs of p are relaxed in the order $(v_0, v_1), (v_1, v_2), \ldots, (v_{k-1}, v_k)$, then $d[v_k] = \delta(v_0, v_k)$. This property holds regardless of any other relaxation steps that occur, even if they are intermixed with relaxations of the edges of p.*

Using these properties, the proof for the optimality of LSP algorithm to find optimal shortest paths in an unconstrained setting is given below.

Lemma 2.6.1 (LSP Optimal for Shortest Path Problem (SPP))**.** *After one complete pass of the LSP algorithm, we have $\delta(v_0, v_j)$ for all vertices $v_j \in \mathcal{V}$ that are reachable from v_0.*

Proof. Consider any vertex v_j that is reachable from v_0 and let $p = v_0, v_1, \ldots, v_j$ be any acyclic shortest path from v_0 to v_j. Path p has at most $|\mathcal{V}| - 1$ arcs, and so $j \leq |\mathcal{V}| - 1$. Each path extension from one layer to the next layer relaxes all network arcs $\mathcal{A}$. Thus, by the path relaxation property, $d[v_j] = \delta(v_0, v_k)$ and the LSP heuristic finds shortest paths. $\square$

Since LSP algorithm can be used to find optimal shortest paths, we compare the performance with the generic Bellman-Ford-Moore algorithm for shortest paths.

2.6.2.1   Comparison of LSP Algorithm and BFM Algorithm

	BFM algorithm	LSP algorithm
Computational Complexity	$\Phi(|\mathcal{V}||\mathcal{A}|)$	$\Phi(|\mathcal{V}|^3)$
Number of Labels stored	$|\mathcal{V}|$	$|\mathcal{V}|^2$

For dense networks, $|\mathcal{A}| \approx |\mathcal{V}|^2$. Therefore, for the dense networks that arise in ESPPRC subproblems for column generation in routing applications, using the BFM algorithm as a heuristic for solving the ESPPRC would require worst-case complexity of $\Phi(|\mathcal{V}|^3)$. Thus, the LSP heuristic generates more candidate routes with similar computational effort for ESPPRC subproblems associated with column generation.

2.7 A Multi-label Extension of LSP Heuristic

2.7.1 k-LSP Heuristic

The k-LSP pricing heuristic extends the LSP method by storing the k smallest reduced-cost routes at each element in the move request array structure described in the previous subsection. This improves the chances of finding more negative reduced cost columns when the approach is used in column generation for routing problems, thereby increasing the solution quality. For sufficiently large k, this method can be shown to find optimal solutions to the ESPPRC.

2.8 Computational Experimentation with Solomon's VRPTW Instances

Computational experiments were conducted to verify the effectiveness of the LSP pricing heuristic in a root column generation heuristic for solving VRPTW problems. The experiments use the benchmark problem sets developed by Solomon (1987) and available at Solomon (2005). These instances vary several characteristics that affect the behavior of routing and scheduling algorithms, including: geographical distribution of customers; the number of customers that can be feasibly served by a vehicle; percent of time-constrained customers; and tightness and positioning of the time windows. The geographical data are randomly generated in problem sets R1 and R2, clustered in problem sets C1 and C2, and a mix of random and clustered structures in problem sets by RC1 and RC2. Problem sets R1, C1 and RC1 have a short scheduling horizon and allow only a few customers per route (approximately 5 to 10). In contrast, the sets R2, C2 and RC2 have a long scheduling horizon permitting many customers (more than 30) to be serviced by the same vehicle. The large problems contain 100 customers, with travel times equal to the Euclidean distance between points. For each such large problem, smaller problems are created by considering only the first 25 or 50 customers.

In the experiments, we use the standard root column generation heuristic to solve the VRPTW, and use the LSP heuristic to solve the pricing subproblems. The approach is implemented in the C programming language, and utilizes the CPLEX Version 8.0 callable libraries for the solution of linear and binary integer programs when necessary. All tests were run on a dual-CPU 2.4 GHz Pentium with 2 GB of memory running Linux. Computation times in the tables to follow are given in seconds. The default branch-and-bound algorithm in CPLEX was allowed a maximum of 30 minutes, and the final solution reported is the best integer solution found in that time period.

2.8.1 Comparison of LSP Heuristic with an Exact ESPPRC Algorithm

The effectiveness of the LSP heuristic approach as a fast, polynomial-time heuristic for the Elementary Shortest Path Problem with Resource Constraints (ESPPRC) subproblem is evaluated first by comparing the solutions obtained for the instances with the results provided in D. Feillet and Gueguen (2004) which are generated by applying an exact algorithm for ESPPRC. In this study, we compare the lower bounds obtained by solving the root linear program using these algorithms within a column generation scheme.

Tables 2 to 19 summarize the comparison results. Columns 2, 3 and 4 report the optimal ESPPRC algorithm solution, if available. Column 2 provides the lower bound obtained by solving the root node linear program using the optimal ESPPRC algorithm, column 3 provides the number of columns generated using the optimal ESPPRC algorithm, and column 4 provides the computation time in seconds required to solve that particular instance. Blank entries in these columns indicate instances with no available solution using the optimal ESPPRC algorithm. Columns 5, 6 and 7 reports the corresponding values using the LSP heuristic. Column 8 reports the percentage gap between the lower bounds obtained using the optimal algorithm and the LSP heuristic. Finally, columns 9 and 10 provide comparative ratios for the number of columns and the computation times for the two algorithms.

Overall, the LSP heuristic solution matched the optimal ESPPRC algorithm solution for 20 out of the 127 instances for which solutions were available in D. Feillet and Gueguen (2004).

The ratio of the computation times for the two algorithms shows that the computation time savings of the heuristic are dramatic. This substantiates the fact that each iteration

Table 2: Comparison of LSP Heuristic and Optimal ESPPRC Algorithm on R1XX.25 instances

Problem Instance	Optimal ESP Algorithm			LSP Heuristic			Solution Gap	$\frac{NC_{LSP}}{NC_{Opt}}$	$\frac{CPU_{LSP}}{CPU_{Opt}}$
	Distance	NC_{Opt}	CPU_{Opt}	Distance	NC_{LSP}	CPU_{LSP}			
R101	617.1	98	0.05	617.1	113	0.01	0	1.153	0.213
R102	546.333	391	0.19	547.1	222	0.01	0.14	0.568	0.053
R103	454.6	710	0.52	469.75	407	0.02	3.23	0.573	0.039
R104	416.9	894	1.13	429.65	566	0.06	2.97	0.633	0.053
R105	530.5	214	0.11	530.5	214	0.03	0	1	0.275
R106	457.3	559	0.36	465.4	353	0.03	1.74	0.631	0.083
R107	424.3	702	0.83	434.8	553	0.04	2.41	0.788	0.048
R108	396.821	1139	1.74	402.082	701	0.1	1.31	0.615	0.058
R109	441.3	360	0.17	441.3	418	0.06	0	1.161	0.351
R110	438.35	676	0.69	444.95	343	0.03	1.48	0.507	0.044
R111	427.283	613	0.66	428.467	474	0.06	0.28	0.773	0.091
R112	387.05	1134	2.2	393	608	0.1	1.51	0.536	0.045

of the LSP heuristic can be performed in polynomial time, and that in contrast the optimal ESPPRC algorithm requires exponential time, Note that the two sets of experiments were conducted using different computational environments. The optimal ESPPRC algorithm was tested on a 1.6-GHz processor with 256 Megabytes of RAM whereas the LSP heuristic was tested on a 2.4 GHz processor with 2 GB of RAM.

The LSP heuristic generally returned high quality solutions for the R1 class, with average solution gaps of less than 1.26%, 2.48% and 3.08% for the 25 customer instances (Table 2), for the 50 customer instances (Table 3) and for the 100 customer instances (Table 4) respectively. For the R2 class, the corresponding average solution gaps were 2.46% (Table 12), 2.59% (Table 12) and 1.62% (Table 13) respectively.

For the C1 (Tables 5 - 7) and C2 (Tables 14 - 16) classes of instances, the overall average solution gap was less than 1%.

For the RC1 and RC2 classes, the solution gaps were slightly higher than the rest of the classes. For the RC1 class, the solution gap was 4.47% for the 25 customer instances (Table 8), 6.32% for the 50 customer instances (Table 9) and 3.9% for the 100 customer instances (Table 10). For the RC2 class, the corresponding solution gaps were 2.12% (Table 17), 3.85% (Table 18) and 3.53% (Table 19) respectively.

Table 3: Comparison of LSP Heuristic and Optimal ESPPRC Algorithm on R1XX.50 instances

Problem Instance	Optimal ESP Algorithm			LSP Heuristic			Solution Gap	$\frac{NC_{LSP}}{NC_{Opt}}$	$\frac{CPU_{LSP}}{CPU_{Opt}}$
	Distance	NC_{Opt}	CPU_{Opt}	Distance	NC_{LSP}	CPU_{LSP}			
R101	1043.37	403	0.42	1043.37	528	0.04	0	1.31	0.095
R102	909	1238	1.61	909.8	885	0.11	0.09	0.715	0.068
R103	769.233	2734	9.89	798.3	1224	0.16	3.64	0.448	0.016
R104	619.077	7222	232	636.295	2736	0.72	2.71	0.379	0.003
R105	892.12	962	1.22	892.767	858	0.15	0.07	0.892	0.123
R106	791.367	1993	4.89	804.435	1324	0.18	1.62	0.664	0.037
R107	707.26	2796	16	735.473	1822	0.32	3.84	0.652	0.02
R108	594.699	6453	338	618.392	3252	0.9	3.83	0.504	0.003
R109	775.342	1511	3.5	788.255	1305	0.23	1.64	0.864	0.066
R110	695.061	2444	8.7	735.245	1553	0.21	5.47	0.635	0.024
R111	696.285	2954	17.8	725.177	1759	0.28	3.98	0.595	0.016
R112	614.851	4110	58.1	633.206	2123	0.49	2.9	0.517	0.008

Table 4: Comparison of LSP Heuristic and Optimal ESPPRC Algorithm on R1XX.100 instances

Problem Instance	Optimal ESP Algorithm			LSP Heuristic			Solution Gap	$\frac{NC_{LSP}}{NC_{Opt}}$	$\frac{CPU_{LSP}}{CPU_{Opt}}$
	Distance	NC_{Opt}	CPU_{Opt}	Distance	NC_{LSP}	CPU_{LSP}			
R101	1631.15	1736	6.58	1631.2	2104	0.4	0	1.212	0.061
R102	1466.6	5419	49.6	1478.25	3645	0.85	0.79	0.673	0.017
R103	1206.78	11800	387	1255.43	5580	1.93	3.88	0.473	0.005
R104				1030.65	7648	2.87			
R105	1346.14	3200	23.4	1355.53	3881	1.37	0.69	1.213	0.059
R106	1226.91	8360	199	1253.56	5041	1.66	2.13	0.603	0.008
R107	1053.45	14198	3561	1109.09	7873	4.53	5.02	0.555	0.001
R108				966.679	10307	6.49			
R109	1134.23	6460	91	1185.96	4849	1.66	4.36	0.751	0.018
R110	1055.57	9878	482	1121.36	6173	2.54	5.87	0.625	0.005
R111	1034.73	14077	975	1089.37	6914	3.16	5.02	0.491	0.003
R112				972.376	9226	6.79			

Table 5: Comparison of LSP Heuristic and Optimal ESPPRC Algorithm on C1XX.25 instances

Problem Instance	Optimal ESP Algorithm			LSP Heuristic			Solution Gap	$\frac{NC_{LSP}}{NC_{Opt}}$	$\frac{CPU_{LSP}}{CPU_{Opt}}$
	Distance	NC_{Opt}	CPU_{Opt}	Distance	NC_{LSP}	CPU_{LSP}			
C101	191.3	647	0.531	191.3	501	0.04	0	0.774	0.075
C102	190.3	1677	2.766	190.3	1303	0.22	0	0.777	0.08
C103	190.3	3598	24.24	196	1983	0.39	2.91	0.551	0.016
C104	186.9	5156	127.5	189	2254	0.91	1.11	0.437	0.007
C105	191.3	880	0.656	191.3	877	0.15	0	0.997	0.229
C106	191.3	663	0.562	191.3	571	0.07	0	0.861	0.125
C107	191.3	932	1.391	191.3	1233	0.27	0	1.323	0.194
C108	191.3	883	1.5	195.1	1948	0.91	1.95	2.206	0.607
C109	191.3	1396	3.391	191.3	1662	0.43	0	1.191	0.127

Table 6: Comparison of LSP Heuristic and Optimal ESPPRC Algorithm on C1XX.50 instances

Problem Instance	Optimal ESP Algorithm			LSP Heuristic			Solution Gap	$\frac{NC_{LSP}}{NC_{Opt}}$	$\frac{CPU_{LSP}}{CPU_{Opt}}$
	Distance	NC_{Opt}	CPU_{Opt}	Distance	NC_{LSP}	CPU_{LSP}			
C101	362.4	1337	2.844	362.4	1455	0.26	0	1.088	0.091
C102	361.4	6980	87.69	362.2	3366	1.05	0.22	0.482	0.012
C103	361.4	6980	87.69	371.1	5920	2.43	2.61	0.848	0.028
C104				376.6	6115	2.98			
C105	362.4	1487	4.515	362.4	2133	0.4	0	1.434	0.089
C106	362.4	1446	2.656	362.4	1792	0.31	0	1.239	0.117
C107	362.4	1737	4.797	362.4	3847	1.28	0	2.215	0.267
C108	362.4	1895	9.843	366.5	4314	1.67	1.12	2.277	0.17
C109	362.4	3443	26.95	365.4	4696	2.09	0.82	1.364	0.078

Table 7: Comparison of LSP Heuristic and Optimal ESPPRC Algorithm on C1XX.100 instances

Problem Instance	Optimal ESP Algorithm			LSP Heuristic			Solution Gap	$\frac{NC_{LSP}}{NC_{Opt}}$	$\frac{CPU_{LSP}}{CPU_{Opt}}$
	Distance	NC_{Opt}	CPU_{Opt}	Distance	NC_{LSP}	CPU_{LSP}			
C101	827.3	2652	18.7	827.3	4802	1.68	0	1.811	0.09
C102	827.3	12315	1066	854.4	11352	9.2	3.17	0.922	0.009
C103				876.6	14685	16.36			
C104				859.2	19113	24.35			
C105	827.3	3828	40.39	827.3	8077	4.77	0	2.11	0.118
C106	827.3	5352	80.36	837.3	9729	6.63	1.2	1.818	0.083
C107	827.3	4055	39.89	827.3	12562	9.95	0	3.098	0.249
C108	827.3	5585	127.8	835.6	13070	15.73	0.99	2.34	0.123
C109	827.3	10508	446	853.6	12690	18.16	3.08	1.208	0.041

Table 8: Comparison of LSP Heuristic and Optimal ESPPRC Algorithm on RC1XX.25 instances

Problem	Optimal ESP Algorithm			LSP Heuristic			Solution	$\frac{NC_{LSP}}{NC_{Opt}}$	$\frac{CPU_{LSP}}{CPU_{Opt}}$
Instance	Distance	NC_{Opt}	CPU_{Opt}	Distance	NC_{LSP}	CPU_{LSP}	Gap		
RC101	406.625	341	0.312	416.207	282	0.03	2.3	0.827	0.096
RC102	351.8	674	0.89	384.7	386	0.04	8.55	0.573	0.045
RC103	332.8	756	1.094	339.2	467	0.07	1.89	0.618	0.064
RC104	306.6	829	2.89	330	690	0.07	7.09	0.832	0.024
RC105	411.3	414	0.546	430.125	346	0.02	4.38	0.836	0.037
RC106	345.5	546	0.75	356.3	390	0.05	3.03	0.714	0.067
RC107	298.3	864	2.484	301.1	650	0.11	0.93	0.752	0.044
RC108	294.5	1269	4.719	318.7	578	0.06	7.59	0.455	0.013

Table 9: Comparison of LSP Heuristic and Optimal ESPPRC Algorithm on RC1XX.50 instances

Problem	Optimal ESP Algorithm			LSP Heuristic			Solution	$\frac{NC_{LSP}}{NC_{Opt}}$	$\frac{CPU_{LSP}}{CPU_{Opt}}$
Instance	Distance	NC_{Opt}	CPU_{Opt}	Distance	NC_{LSP}	CPU_{LSP}	Gap		
RC101	850.021	900	1.156	880.918	724	0.1	3.51	0.804	0.087
RC102	721.815	1746	5.812	741.138	1048	0.12	2.61	0.6	0.021
RC103	645.281	2195	25.91	667.311	1566	0.27	3.3	0.713	0.01
RC104	545.8	3378	128.1	597.88	2234	0.41	8.71	0.661	0.003
RC105	761.558	1331	3.703	789.483	1024	0.16	3.54	0.769	0.043
RC106	664.433	1671	5.672	707.184	1116	0.17	6.05	0.668	0.03
RC107	603.583	2338	25.36	669.414	1620	0.49	9.83	0.693	0.019
RC108	541.167	3456	252.5	621.877	1500	0.33	13	0.434	0.001

Table 10: Comparison of LSP Heuristic and Optimal ESPPRC Algorithm on RC1XX.100 instances

Problem	Optimal ESP Algorithm			LSP Heuristic			Solution	$\frac{NC_{LSP}}{NC_{Opt}}$	$\frac{CPU_{LSP}}{CPU_{Opt}}$
Instance	Distance	NC_{Opt}	CPU_{Opt}	Distance	NC_{LSP}	CPU_{LSP}	Gap		
RC101	1584.09	2829	16.89	1599.39	2776	0.88	0.96	0.981	0.052
RC102	1406.26	5633	87.67	1454.21	3767	0.98	3.3	0.669	0.011
RC103	1225.65	10217	800.9	1303.49	5507	2.14	5.97	0.539	0.003
RC104				1179.28	8713	5.19			
RC105	1471.92	4809	51.13	1514.14	3485	0.94	2.79	0.725	0.018
RC106	1318.8	5172	82.95	1367.66	4233	1.82	3.57	0.818	0.022
RC107	1183.37	8491	535.7	1269.98	5474	3.03	6.82	0.645	0.006
RC108				1165.22	6575	3.27			

Table 11: Comparison of LSP Heuristic and Optimal ESPPRC Algorithm on R2XX.25 instances

Problem	Optimal ESP Algorithm			LSP Heuristic			Solution	$\frac{NC_{LSP}}{NC_{Opt}}$	$\frac{CPU_{LSP}}{CPU_{Opt}}$
Instance	Distance	NC_{Opt}	CPU_{Opt}	Distance	NC_{LSP}	CPU_{LSP}	Gap		
R201	460.1	569	0.703	460.1	968	0.11	0	1.701	0.156
R202	410.5	1384	2.297	413	1194	0.12	0.61	0.863	0.052
R203	391.4	1985	5.516	396.75	1434	0.23	1.35	0.722	0.042
R204	350.475	5941	89.88	368.45	2887	0.68	4.88	0.486	0.008
R205	390.6	1332	2.843	398.813	1296	0.17	2.06	0.973	0.06
R206	373.6	2808	9.734	385.6	1890	0.35	3.11	0.673	0.036
R207	360.05	3893	24.58	367.525	2155	0.27	2.03	0.554	0.011
R208	328.2	8237	310.7	337.444	3206	0.81	2.74	0.389	0.003
R209	364.05	2514	7.735	370.25	1763	0.34	1.67	0.701	0.044
R210	404.175	1875	3.547	422.035	1545	0.2	4.23	0.824	0.056
R211	341.327	4653	41.38	356.797	1930	0.38	4.34	0.415	0.009

Table 12: Comparison of LSP Heuristic and Optimal ESPPRC Algorithm on R2XX.50 instances

Problem	Optimal ESP Algorithm			LSP Heuristic			Solution	$\frac{NC_{LSP}}{NC_{Opt}}$	$\frac{CPU_{LSP}}{CPU_{Opt}}$
Instance	Distance	NC_{Opt}	CPU_{Opt}	Distance	NC_{LSP}	CPU_{LSP}	Gap		
R201	791.9	2639	9	794.98	5913	2.03	0.39	2.241	0.226
R202	698.5	7747	70.64	712.96	6239	1.8	2.03	0.805	0.025
R203	598.583	14984	538.3	634.325	8871	3.13	5.63	0.592	0.006
R204				522.562	21803	23.39			
R205	682.85	5759	38.03	696.944	6987	2.83	2.02	1.213	0.074
R206	626.343	11653	255	651.964	9797	4.2	3.93	0.841	0.016
R207				589.708	12851	8.11			
R208				495.044	28552	33.73			
R209	599.825	8066	174.7	608	8934	4.72	1.34	1.108	0.027
R210	636.1	10771	263.7	654.15	9843	6.76	2.76	0.914	0.026
R211				549.251	8884	4.47			

Table 13: Comparison of LSP Heuristic and Optimal ESPPRC Algorithm on R2XX.100 instances

Problem Instance	Optimal ESP Algorithm			LSP Heuristic			Solution Gap	$\frac{NC_{LSP}}{NC_{Opt}}$	$\frac{CPU_{LSP}}{CPU_{Opt}}$
	Distance	NC_{Opt}	CPU_{Opt}	Distance	NC_{LSP}	CPU_{LSP}			
R201	1140.3	10101	205.9	1156.05	23174	21.54	1.36	2.294	0.105
R202				1045.72	37954	60.46			
R203				921.83	52907	70.78			
R204				771.727	106874	299.2			
R205	939.124	27744	2763	957.11	43277	84.58	1.88	1.56	0.031
R206				904.405	52648	91.24			
R207				831.43	73126	181.8			
R208				727.767	133366	485.6			
R209				879.21	44090	76.24			
R210				931.831	46546	74.06			
R211				773.893	55565	114.2			

Table 14: Comparison of LSP Heuristic and Optimal ESPPRC Algorithm on C2XX.25 instances

Problem Instance	Optimal ESP Algorithm			LSP Heuristic			Solution Gap	$\frac{NC_{LSP}}{NC_{Opt}}$	$\frac{CPU_{LSP}}{CPU_{Opt}}$
	Distance	NC_{Opt}	CPU_{Opt}	Distance	NC_{LSP}	CPU_{LSP}			
C201	214.7	1326	1.891	214.7	781	60.08	0	0.589	31.77
C202	214.7	3708	18.83	214.7	2044	107.6	0	0.551	5.714
C203	214.7	6439	117.8	218.4	3506	109.6	1.69	0.544	0.93
C204				227.3	6042	111.9			
C205	214.7	3266	12.61	219.9	1684	80.19	2.36	0.516	6.36
C206	214.7	3477	14.23	216.6	2584	83.35	0.88	0.743	5.856
C207	214.5	6292	64.88	214.5	2973	87.44	0	0.473	1.348
C208	214.5	5002	29.84	216	3453	80.3	0.69	0.69	2.691

Table 15: Comparison of LSP Heuristic and Optimal ESPPRC Algorithm on C2XX.50 instances

Problem Instance	Optimal ESP Algorithm			LSP Heuristic			Solution Gap	$\frac{NC_{LSP}}{NC_{Opt}}$	$\frac{CPU_{LSP}}{CPU_{Opt}}$
	Distance	NC_{Opt}	CPU_{Opt}	Distance	NC_{LSP}	CPU_{LSP}			
C201	360.2	9443	130.9	360.2	4136	188	0	0.438	1.437
C202				388.6	13629	389.4			
C203				389.02	19610	350.2			
C204				392.77	31426	419			
C205	359.8	20002	758.7	368.1	25310	238.8	2.25	1.265	0.315
C206	359.8	30285	1690	369.1	19940	191.7	2.52	0.658	0.113
C207				371.8	36519	216.1			
C208				359.4	28048	200.3			

Table 16: Comparison of LSP Heuristic and Optimal ESPPRC Algorithm on C2XX.100 instances

Problem Instance	Optimal ESP Algorithm			LSP Heuristic			Solution Gap	$\frac{NC_{LSP}}{NC_{Opt}}$	$\frac{CPU_{LSP}}{CPU_{Opt}}$
	Distance	NC_{Opt}	CPU_{Opt}	Distance	NC_{LSP}	CPU_{LSP}			
C201	589.1	24492	1121	589.1	23323	530.1	0	0.952	0.473
C202				616	78879	499.2			
C203				677.46	63146	830.9			
C204				671.95	88181	928.2			
C205				625.5	60825	506.9			
C206				627.42	56455	427.7			
C207				639.33	57157	524.4			
C208				628.62	76925	415.8			

Table 17: Comparison of LSP Heuristic and Optimal ESPPRC Algorithm on RC2XX.25 instances

Problem	Optimal ESP Algorithm			LSP Heuristic			Solution	$\frac{NC_{LSP}}{NC_{Opt}}$	$\frac{CPU_{LSP}}{CPU_{Opt}}$
Instance	Distance	NC_{Opt}	CPU_{Opt}	Distance	NC_{LSP}	CPU_{LSP}	Gap		
RC201	360.2	770	0.562	360.6	767	0.13	0.11	0.996	0.231
RC202	338	1791	3.953	338.8	1046	0.08	0.24	0.584	0.02
RC203	326.9	4547	63.13	336.4	1434	0.18	2.82	0.315	0.003
RC204	299.7	6728	121.4	314.5	2003	0.36	4.71	0.298	0.003
RC205	338	1045	1.453	345.9	886	0.08	2.28	0.848	0.055
RC206	324	1065	1.484	334.6	1530	0.32	3.17	1.437	0.216
RC207	298.3	1474	5.531	302.9	1640	0.2	1.52	1.113	0.036
RC208				284.7	3109	1.83			

Table 18: Comparison of LSP Heuristic and Optimal ESPPRC Algorithm on RC2XX.50 instances

Problem	Optimal ESP Algorithm			LSP Heuristic			Solution	$\frac{NC_{LSP}}{NC_{Opt}}$	$\frac{CPU_{LSP}}{CPU_{Opt}}$
Instance	Distance	NC_{Opt}	CPU_{Opt}	Distance	NC_{LSP}	CPU_{LSP}	Gap		
RC201	684.8	2584	8.312	702.4	4218	1.09	2.51	1.632	0.131
RC202	613.6	6245	46.8	637.7	5925	1.53	3.78	0.949	0.033
RC203				617.8	7990	3.29			
RC204				502.1	29583	33.38			
RC205	630.2	4749	28.08	662	5155	1.42	4.8	1.085	0.051
RC206	610	5272	31.08	637.4	6825	2.27	4.3	1.295	0.073
RC207				593.3	10552	4.79			
RC208				520.855	14635	16.69			

2.8.2 Comparison of LSP Heuristic Solutions to Best Known Solutions

The quality of the solutions generated by a root column generation heuristic using the LSP heuristic for the pricing subproblem is now evaluated by comparing its solutions for Solomon's VRPTW instances to the optimal solutions (if known) to those problems and the best heuristic solutions for the 100 customer instances of these problems, which are reported at Solomon (2005). Tables 20 to 37 summarize the comparison of the root column generation heuristic using LSP with the actual optimal IP solution on these instances. Columns 2 and 3 report the optimal solution, if available. Column 3 provides the optimal travel distance, and column 2 provides the number of vehicles used to obtain that travel distance. Blank entries in columns 2 and 3 represent instances that to date have not been solved to optimality. The

Table 19: Comparison of LSP Heuristic and Optimal ESPPRC Algorithm on RC2XX.100 instances

Problem	Optimal ESP Algorithm			LSP Heuristic			Solution	$\frac{NC_{LSP}}{NC_{Opt}}$	$\frac{CPU_{LSP}}{CPU_{Opt}}$
Instance	Distance	NC_{Opt}	CPU_{Opt}	Distance	NC_{LSP}	CPU_{LSP}	Gap		
RC201	1255.94	10031	282.1	1270.99	27060	34.13	1.18	2.698	0.121
RC202	1088.08	31972	2411	1138.55	32388	52.09	4.43	1.013	0.022
RC203				999.487	54347	94.29			
RC204				865.208	91259	199.2			
RC205	1147.61	20099	1492	1207.61	26859	25.75	4.97	1.336	0.017
RC206				1062.92	35816	78.87			
RC207				1007.41	39598	55.23			
RC208				825.682	53737	130.6			

third, fourth and fifth columns report the LSP solution. Column 3 provides the heuristic objective function value of the root node linear program solved using heuristic column generation with LSP pricing; column 4 provides the final integer solution after solving the branch-and-bound using the default CPLEX approach; and column 4 provides the number of used vehicles in this solution. The sixth and seventh columns report the average CPU execution time in seconds required to solve the the heuristic column generation and the branch-and-bound for the final integer program. Column 8 provides the percentage gap between the final integer solution and the root node objective; and column 9 provides the gap between the final heuristic solution and the optimal solution for each instance. Rows represented in bold indicate instances where the LSP heuristic finds the optimal solution.

The LSP heuristic finds the optimal solution for 24 out of the 168 instances, of which 6 belonged to the R1XX.25 category (Table 20), 6 to the C1XX.25 category (Table 23), 4 to the C1XX.50 category (Table 24), 3 to the C1XX.100 category (Table 25), 3 to the C2XX.25 category (Table 32), and one each from the C2XX.50 category (Table 24) and the C2XX.100 category (Table 34)

The gap between the final integer solution and the lower bound obtained at root node after heuristic column generation was less than 5% for most of the instances that were solved to completion within the 30 minute branch-and-bound time limit for the R and C classes. For instances solved to optimality using the LSP heuristic, an optimal integer solution was

34

always obtained directly by solving the root LP. Such results indicate that good results can be obtained by using heuristic column generation only at the root node, instead of adopting a branch-and-price scheme.

The heuristic returned high quality solutions for the R1 class, with average optimality gaps of less than 2% for the 25 customer instances (Table 20), less than 5% for the 50 customer instances (Table 21) and less than 8% for the 100 customer instances (Table 22). For instances with optimality gaps greater than 10%, *e.g.*, R104.100 and R111.100, the gap between the IP solution and the LP objective was also higher due to the 30 minute time restriction on the branch-and-bound. This heuristic LP lower bound obtained is still tight and close to the actual optimal value, suggesting that the heuristic performance is consistent across various instances in the R1 class. Also, the CPU time required for solution of the root node linear program via heuristic column generation was less than a second for all the 25 and 50 customer instances, and the 100 customer instances were solved within 10 seconds.

The heuristic produced very high quality solutions for the C1 class, with average optimality gaps of less than 1% for the 25 customer instances (Table 23), less than 2% for the 50 customer instances (Table 24) and less than 4% for the 100 customer instances (Table 25). Also, the CPU time required for the root node column generation averaged less than a second for the 25 customer instances, and 50 customer instances averaged 2 seconds, and the 100 customer instances averaged close to 12 seconds.

For the RC1 class, the optimality gaps were higher, with average gaps of 7.9%; the maximum gap was 14.5% for RC102.25. The heuristic column generation was performed within a second for the 25 and 50 customer instances, and averaged around 2 seconds for the 100 customer instances.

The heuristic returned good quality solutions for the R2 class as well, which have long scheduling horizon permitting many customers (more than 30) to be serviced by the same vehicle. The performance, however, was not as good as that for the R1 class. The average optimality gaps were less than 4% for the 25 customer instances (Table 29), less than 8% for the 50 customer instances (Table 30) and less than 7% for the 100 customer instances (Table

35

31). Similar to the R1 class, optimality gaps of more than 10% (for example R204.50) can be attributed to the fact that the branch-and-bound tree was truncated with the 30 minute limit. The lower bound obtained is still tight and close to the actual optimal value. The CPU time for the heuristic was less than a second for all the 25 instances, and 50 customer instances were solved within 10 seconds and the 100 customer instances took about 150 seconds.

For the C2 class, the performance was adequate but worse than that for the C1 class. Average optimality gaps were less than 2% for the 25 customer instances (Table 32), less than 7% for the 50 customer instances (Table 33) and less than 13% for the 100 customer instances (Table 34). Again, most of the higher optimality gaps were a result of early termination of the branch-and-bound tree. The CPU time required for all the 25 customer instances where averaged less than 2 seconds and 50 customer instances averaged 40 seconds, and the 100 customer instances averaged close to 220 seconds.

For the RC2 class, the optimality gaps were slightly lower than those for RC1, with the average about 5% with a maximum of 12% for RC204.50. The heuristic CPU time averaged less than a second for the 25 customer instances, around 8 seconds for 50 customer instances and around 80 seconds for the 100 customer instances.

Tables 38 to 43 summarize the results of a comparison of the root column generation heuristic using LSP with the best heuristic solution on these instances. Columns 2 and 3 report the best heuristic solution. Column 3 provides the travel distance, and column 2 provides the number of vehicles used to obtain that travel distance. The fourth and fifth columns report the LSP solution. Column 5 provides the final integer solution using LSP heuristic after solving the branch-and-bound using the default CPLEX approach; and column 4 provides the number of used vehicles in this solution. Column 6 provides the gap between the best heuristic solution and the LSP heuristic solution for each instance. The LSP heuristic found solutions better than existing best heuristic solutions for 12 out of the 56 100 customer instances, showing that it is a very useful heuristic for generating fast, high-quality solutions to routing problems with resource constraints.

Table 20: Performance of LSP Heuristic on R1XX.25 instances

Problem	Optimal Solution		LSP Heuristic Solution			Time		$\frac{IP-LP}{IP}$	Opt
Instance	NV	Distance	LP	IP	NV	Heuristic	B & B		Gap
R101	**8**	**617.1**	**617.1**	**617.1**	**8**	**0.01**	**0**	**0**	**0**
R102	**7**	**547.1**	**547.1**	**547.1**	**7**	**0.01**	**0**	**0**	**0**
R103	5	454.6	469.75	478.4	6	0.02	0.02	0.018	4.97
R104	4	416.9	429.65	430.2	4	0.06	0.02	0.001	3.09
R105	**6**	**530.5**	**530.5**	**530.5**	**6**	**0.03**	**0**	**0**	**0**
R106	**3**	**465.4**	**465.4**	**465.4**	**5**	**0.03**	**0.01**	**0**	**0**
R107	4	424.3	434.8	447.9	5	0.04	0.11	0.029	5.27
R108	4	397.3	402.082	407.2	4	0.1	0.04	0.013	2.43
R109	**5**	**441.3**	**441.3**	**441.3**	**5**	**0.06**	**0**	**0**	**0**
R110	4	444.1	444.95	445.7	4	0.03	0.01	0.002	0.36
R111	5	428.8	428.467	430.1	5	0.06	0.01	0.004	0.3
R112	**4**	**393**	**393**	**393**	**4**	**0.1**	**0.01**	**0**	**0**

Table 21: Performance of LSP Heuristic on R1XX.50 instances

Problem	Optimal Solution		LSP Heuristic Solution			Time		$\frac{IP-LP}{IP}$	Opt
Instance	NV	Distance	LP	IP	NV	Heuristic	B & B		Gap
R101	12	1044	1043.37	1045.7	12	0.04	0.01	0.002	0.16
R102	11	909	909.8	909.8	11	0.11	0.01	0	0.09
R103	9	772.9	798.3	802.9	9	0.16	0.05	0.006	3.74
R104	6	625.4	636.295	657.4	6	0.72	3.8	0.032	4.87
R105	9	899.3	892.767	908.5	10	0.15	0.46	0.017	1.01
R106	5	793	804.435	813.2	8	0.18	0.24	0.011	2.48
R107	7	711.1	735.473	758.1	8	0.32	4.18	0.03	6.2
R108	6	617.7	618.392	655.4	6	0.9	20.97	0.056	5.75
R109	8	786.8	788.255	806.9	9	0.23	0.44	0.023	2.49
R110	7	697	735.245	762.1	8	0.21	2.08	0.035	8.54
R111	7	707.2	725.177	767.9	8	0.28	9.07	0.056	7.9
R112	6	630.2	633.206	682.2	7	0.49	14.52	0.072	7.62

Table 22: Performance of LSP Heuristic on R1XX.100 instances

Problem	Optimal Solution		LSP Heuristic Solution			Time		$\frac{IP-LP}{IP}$	Opt
Instance	NV	Distance	LP	IP	NV	Heuristic	B & B		Gap
R101	20	1637.7	1631.2	1639.8	20	0.4	0.29	0.005	0.13
R102	18	1466.6	1478.25	1483.6	18	0.85	3.73	0.0036	1.15
R103	14	1208.7	1255.43	1293	15	1.93	1146.66	0.03	6.52
R104	11	971.5	1030.65	1138.7	13	2.87	1800.0	0.095	14.7
R105	15	1355.3	1355.53	1382.9	16	1.37	44.2	0.02	2
R106	13	1234.6	1253.56	1322	14	1.66	1800.0	0.052	6.61
R107	11	1064.6	1109.09	1219.7	14	4.53	1800.0	0.091	12.7
R108			966.679	1048.5	11	6.49	1800.0	0.078	
R109	13	1146.9	1185.96	1245	14	1.66	1800.0	0.047	7.88
R110	12	1068	1121.36	1178.2	13	2.54	1367.11	0.0482	9.35
R111	12	1048.7	1089.37	1239.5	14	3.16	1800.0	0.121	15.4
R112			972.376	1109.9	13	6.79	1800.0	0.124	

Table 23: Performance of LSP Heuristic on C1XX.25 instances

Problem	Optimal Solution		LSP Heuristic Solution			Time		$\frac{IP-LP}{IP}$	Opt
Instance	NV	Distance	LP	IP	NV	Heuristic	B & B		Gap
C101	**3**	**191.3**	**191.3**	**191.3**	**3**	**0.04**	**0**	**0**	**0**
C102	**3**	**190.3**	**190.3**	**190.3**	**3**	**0.22**	**0.04**	**0**	**0**
C103	3	190.3	196	196	3	0.39	0.05	0	2.91
C104	3	186.9	189	189	3	0.91	0.07	0	1.11
C105	**3**	**191.3**	**191.3**	**191.3**	**3**	**0.15**	**0.01**	**0**	**0**
C106	**3**	**191.3**	**191.3**	**191.3**	**3**	**0.07**	**0.01**	**0**	**0**
C107	**3**	**191.3**	**191.3**	**191.3**	**3**	**0.27**	**0.01**	**0**	**0**
C108	3	191.3	195.1	195.1	3	0.91	0.06	0	1.95
C109	**3**	**191.3**	**191.3**	**191.3**	**3**	**0.43**	**0.05**	**0**	**0**

Table 24: Performance of LSP Heuristic on C1XX.50 instances

Problem	Optimal Solution		LSP Heuristic Solution			Time		$\frac{IP-LP}{IP}$	Opt
Instance	NV	Distance	LP	IP	NV	Heuristic	B & B		Gap
C101	**5**	**362.4**	**362.4**	**362.4**	**5**	**0.26**	**0.02**	**0**	**0**
C102	5	361.4	362.2	362.2	5	1.05	0.1	0	0.22
C103	5	361.4	371.1	371.1	5	2.43	0.07	0	2.61
C104	5	358	376.64	399.3	6	2.98	9.35	0.057	10.3
C105	**5**	**362.4**	**362.4**	**362.4**	**5**	**0.4**	**0.02**	**0**	**0**
C106	**5**	**362.4**	**362.4**	**362.4**	**5**	**0.31**	**0.01**	**0**	**0**
C107	**5**	**362.4**	**362.4**	**362.4**	**5**	**1.28**	**0.04**	**0**	**0**
C108	5	362.4	366.5	366.5	5	1.67	0.13	0	1.12
C109	5	362.4	365.4	365.4	5	2.09	0.16	0	0.82

Table 25: Performance of LSP Heuristic on C1XX.100 instances

Problem	Optimal Solution		LSP Heuristic Solution			Time		$\frac{IP-LP}{IP}$	Opt
Instance	NV	Distance	LP	IP	NV	Heuristic	B & B		Gap
C101	**10**	**827.3**	**827.3**	**827.3**	**10**	**1.68**	**0.15**	**0**	**0**
C102	10	827.3	854.36	872.6	10	9.2	10.96	0.021	5.19
C103	10	826.3	876.63	898.4	11	16.36	15.42	0.024	8.03
C104	10	822.9	859.2	859.2	10	24.35	0.18	0	4.22
C105	**10**	**827.3**	**827.3**	**827.3**	**10**	**4.77**	**0.34**	**0**	**0**
C106	10	827.3	837.33	850.4	10	6.63	1.41	0.015	2.72
C107	**10**	**827.3**	**827.3**	**827.3**	**10**	**9.95**	**0.32**	**0**	**0**
C108	10	827.3	835.6	863.2	11	15.73	12.24	0.032	4.16
C109	10	827.3	853.63	860	10	18.16	2.41	0.007	3.8

Table 26: Performance of LSP Heuristic on RC1XX.25 instances

Problem	Optimal Solution		LSP Heuristic Solution			Time		$\frac{IP-LP}{IP}$	Opt
Instance	NV	Distance	LP	IP	NV	Heuristic	B & B		Gap
RC101	4	461.1	416.207	477	5	0.03	0.12	0.127	3.33
RC102	3	351.8	384.7	411.3	4	0.04	0.03	0.065	14.5
RC103	3	332.8	339.2	339.2	3	0.07	0.01	0	1.89
RC104	3	306.6	330	330	3	0.07	0.01	0	7.09
RC105	4	411.3	430.125	453	4	0.02	0.04	0.05	9.21
RC106	3	345.5	356.3	356.3	3	0.05	0	0	3.03
RC107	3	298.3	301.1	301.1	3	0.11	0.01	0	0.93
RC108	3	294.5	318.7	318.7	3	0.06	0.01	0	7.59

Table 27: Performance of LSP Heuristic on RC1XX.50 instances

Problem	Optimal Solution		LSP Heuristic Solution			Time		$\frac{IP-LP}{IP}$	Opt
Instance	NV	Distance	LP	IP	NV	Heuristic	B & B		Gap
RC101	8	944	880.918	1002.2	9	0.1	24.02	0.121	5.81
RC102	7	822.5	741.138	868.9	8	0.12	17.48	0.147	5.34
RC103	6	710.9	667.311	747.2	6	0.27	0.6	0.107	4.86
RC104	5	545.8	597.88	600.2	5	0.41	0.08	0	9.06
RC105	8	855.3	789.483	930.6	8	0.16	21.13	0.152	8.09
RC106	6	723.2	707.184	836	8	0.17	10.36	0.154	13.5
RC107	6	642.7	669.414	714	6	0.49	0.17	0.062	9.99
RC108	6	598.1	621.877	694.5	6	0.33	0.44	0.105	13.9

Table 28: Performance of LSP Heuristic on RC1XX.100 instances

Problem	Optimal Solution		LSP Heuristic Solution			Time		$\frac{IP-LP}{IP}$	Opt
Instance	NV	Distance	LP	IP	NV	Heuristic	B & B		Gap
RC101	15	1619.8	1599.39	1667.8	16	0.88	99.18	0.041	2.88
RC102	14	1457.4	1454.21	1555.1	15	0.98	621.3	0.065	6.28
RC103	11	1258	1303.49	1411.2	13	2.14	1800.0	0.0763	10.9
RC104			1179.28	1294.1	12	5.19	1800.0	0.089	
RC105	15	1513.7	1514.14	1641	17	0.94	1800.0	0.08	7.76
RC106			1367.66	1479.2	14	1.82	1800.0	0.075	
RC107	12	1207.8	1269.98	1359.2	13	3.03	1800.0	0.066	11.1
RC108	11	1114.2	1165.22	1310.2	13	3.27	1800.0	0.111	15

Table 29: Performance of LSP Heuristic on R2XX.25 instances

Problem	Optimal Solution		LSP Heuristic Solution			Time		$\frac{IP-LP}{IP}$	Opt
Instance	NV	Distance	LP	IP	NV	Heuristic	B & B		Gap
R201	4	463.3	460.1	464.7	4	0.11	0.04	0.01	0.3
R202	4	410.5	413	413	4	0.12	0.03	0	0.61
R203	3	391.4	396.75	401	3	0.23	0.07	0.011	2.39
R204	2	355	368.45	373.6	3	0.68	0.25	0.014	4.98
R205	3	393	398.81	404.5	3	0.17	0.08	0.014	2.84
R206	3	374.4	385.6	385.6	3	0.35	0.03	0	2.9
R207	3	361.6	367.53	380.9	3	0.27	0.45	0.035	5.07
R208	1	328.2	337.44	339.7	2	0.81	0.34	0.007	3.39
R209	2	370.7	370.25	376.4	3	0.34	0.1	0.016	1.51
R210	3	404.6	422.04	429.8	4	0.2	0.17	0.018	5.86
R211	2	350.9	356.8	390	4	0.38	4.75	0.085	10

Table 30: Performance of LSP Heuristic on R2XX.50 instances

Problem	Optimal Solution		LSP Heuristic Solution			Time		$\frac{IP-LP}{IP}$	Opt
Instance	NV	Distance	LP	IP	NV	Heuristic	B & B		Gap
R201	6	791.9	794.98	809.2	6	2.03	3.22	0.018	2.14
R202	5	698.5	712.96	732.7	6	1.8	13.22	0.027	4.67
R203	5	605.3	634.33	656.1	5	3.13	40.32	0.033	7.74
R204	2	506.4	522.56	599.2	5	23.39	1800.0	0.128	15.5
R205	4	690.1	696.94	723.1	7	2.83	25.31	0.036	4.56
R206	4	632.4	651.96	686.3	6	4.2	209.25	0.05	7.85
R207			589.71	659.5	5	8.11	1837	0.106	
R208			495.04	550.5	3	33.73	1800.0	0.101	
R209	4	600.6	608	619.9	4	4.72	1.04	0.019	3.11
R210	4	645.6	654.15	683.2	5	6.76	104.69	0.043	5.5
R211	3	535.5	549.25	623	6	4.47	1800.0	0.118	14

Table 31: Performance of LSP Heuristic on R2XX.100 instances

Problem	Optimal Solution		LSP Heuristic Solution			Time		$\frac{IP-LP}{IP}$	Opt
Instance	NV	Distance	LP	IP	NV	Heuristic	B & B		Gap
R201	8	1143.2	1156.1	1222.5	11	21.54	1800.0	0.054	6.49
R202			1045.7	1078.2	9	60.46	1800.0	0.03	
R203			921.83	1049.3	8	70.78	1800.0	0.121	
R204			771.73	908.5	7	299.2	1800.0	0.151	
R205			957.11	1105.9	7	84.58	1800.0	0.135	
R206			904.41	1060.5	8	91.24	1800.0	0.147	
R207			831.43	962.5	8	181.8	1800.0	0.136	
R208			727.77	991.3	6	485.6	1800.0	0.266	
R209			879.21	994.8	7	76.24	1800.0	0.116	
R210			931.83	1004.5	8	74.06	1800.0	0.072	
R211			773.89	920.5	8	114.2	1800.0	0.159	

Table 32: Performance of LSP Heuristic on C2XX.25 instances

| Problem | Optimal Solution | | LSP Heuristic Solution | | | Time | | $\frac{IP-LP}{IP}$ | Opt |
Instance	NV	Distance	LP	IP	NV	Heuristic	B & B		Gap
C201	**2**	**214.7**	**214.7**	**214.7**	**2**	**0.13**	**0**	**0**	**0**
C202	**2**	**214.7**	**214.7**	**214.7**	**2**	**0.42**	**0.07**	**0**	**0**
C203	2	214.7	218.4	218.4	2	1.16	0.03	0	1.69
C204	2	213.1	227.3	227.4	2	3.57	0.11	¡0.001	6.29
C205	2	214.7	219.9	219.9	2	0.34	0.04	0	2.36
C206	2	214.7	216.6	216.6	2	0.91	0.03	0	0.88
C207	**2**	**214.5**	**214.5**	**214.5**	**2**	**1**	**0.02**	**0**	**0**
C208	2	214.5	216	216	2	1.69	0.03	0	0.69

Table 33: Performance of LSP Heuristic on C2XX.50 instances

| Problem | Optimal Solution | | LSP Heuristic Solution | | | Time | | $\frac{IP-LP}{IP}$ | Opt |
Instance	NV	Distance	LP	IP	NV	Heuristic	B & B		Gap
C201	**3**	**360.2**	**360.2**	**360.2**	**3**	**1.19**	**0.05**	**0**	**0**
C202	3	360.2	388.6	395.8	3	7.48	26.13	0.018	8.99
C203	3	359.8	389.02	428	4	16.73	257.69	0.091	15.9
C204	2	350.1	392.77	427.9	3	42.01	279.14	0.082	18.2
C205	3	359.8	368.1	368.1	3	44.48	0.3	0	2.25
C206	3	359.8	369.1	369.1	3	37.75	0.25	0	2.52
C207	3	359.6	371.8	371.8	3	97.53	0.5	0	3.28
C208	2	350.5	359.4	359.4	2	72.53	0.36	0	2.48

Table 34: Performance of LSP Heuristic on C2XX.100 instances

| Problem | Optimal Solution | | LSP Heuristic Solution | | | Time | | $\frac{IP-LP}{IP}$ | Opt |
Instance	NV	Distance	LP	IP	NV	Heuristic	B & B		Gap
C201	**3**	**589.1**	**589.1**	**589.1**	**3**	**20.69**	**0.3**	**0**	**0**
C202	3	589.1	616	634.9	4	384.27	26.76	0.03	7.21
C203	3	588.7	677.46	786.2	5	135.35	1800.0	0.138	25.1
C204	3	588.1	671.95	815.8	5	258.29	1800.0	0.176	27.9
C205	3	586.4	625.5	625.5	3	197.74	1.13	0	6.25
C206	3	586	627.42	648.6	5	211.8	209.04	0.033	9.65
C207	3	585.8	639.33	699.8	5	170.91	1800.0	0.086	16.3
C208	3	585.8	628.62	661.8	4	394.56	720.24	0.05	11.5

Table 35: Performance of LSP Heuristic on RC2XX.25 instances

| Problem | Optimal Solution | | LSP Heuristic Solution | | | Time | | $\frac{IP-LP}{IP}$ | Opt |
Instance	NV	Distance	LP	IP	NV	Heuristic	B & B		Gap
RC201	3	360.2	360.6	360.6	3	0.13	0.01	0	0.11
RC202	3	338	338.8	338.8	3	0.08	0.02	0	0.24
RC203	3	326.9	336.4	336.4	3	0.18	0.04	0	2.82
RC204	3	299.7	314.5	314.5	3	0.36	0.05	0	4.71
RC205	3	338	345.9	345.9	3	0.08	0.02	0	2.28
RC206	3	324	334.6	334.6	3	0.32	0.03	0	3.17
RC207	3	298.3	302.9	302.9	3	0.2	0.04	0	1.52
RC208	2	269.1	284.7	284.7	2	1.83	0.05	0	5.48

Table 36: Performance of LSP Heuristic on RC2XX.50 instances

| Problem | Optimal Solution | | LSP Heuristic Solution | | | Time | | $\frac{IP-LP}{IP}$ | Opt |
Instance	NV	Distance	LP	IP	NV	Heuristic	B & B		Gap
RC201	5	684.8	702.4	702.4	5	1.09	0.13	0	2.506
RC202	5	613.6	637.7	637.7	5	1.53	0.04	0	3.779
RC203	4	555.3	617.8	617.8	5	3.29	0.07	0	10.12
RC204	3	444.2	502.1	502.1	3	33.38	0.41	0	11.53
RC205	5	630.2	662	662	5	1.42	0.04	0	4.804
RC206	5	610	637.4	637.4	5	2.27	0.05	0	4.299
RC207	4	558.6	593.3	593.3	5	4.79	0.09	0	5.849
RC208			520.85	531.9	5	16.69	7.14	0.0208	

Table 37: Performance of LSP Heuristic on RC2XX.100 instances

| Problem | Optimal Solution | | LSP Heuristic Solution | | | Time | | $\frac{IP-LP}{IP}$ | Opt |
Instance	NV	Distance	LP	IP	NV	Heuristic	B & B		Gap
RC201	9	1261.8	1271	1290.9	8	34.13	81.01	0.0154	2.254
RC202	8	1092.3	1138.6	1158	8	52.09	287.94	0.0168	5.674
RC203			999.49	1115.8	8	94.29	1800.0	0.1042	
RC204			865.21	1016	8	199.17	1800.0	0.1484	
RC205	7	1154	1207.6	1282.6	11	25.75	1800.0	0.0585	10.03
RC206			1062.9	1149.5	8	78.87	1800.0	0.0753	
RC207			1007.4	1090.9	8	55.23	1800.0	0.0765	
RC208			825.68	904.7	6	130.58	1800.0	0.0873	

Table 38: Comparison of LSP Heuristic Solution with Best Heuristic Solution: R1XX.100 instances

Problem	Best Heuristic Solution		LSP Heuristic Solution		Solution
Instance	NV	Distance	NV	Distance	Gap %
R101	19	1645.79	20	1639.8	-0.37
R102	17	1486.12	18	1483.6	-0.17
R103	13	1292.68	15	1293	0.025
R104	9	1007.24	13	1138.7	11.54
R105	14	1377.11	16	1382.9	0.419
R106	12	1251.98	14	1322	5.297
R107	10	1104.66	14	1219.7	9.432
R108	9	960.88	11	1048.5	8.357
R109	11	1194.73	14	1245	4.038
R110	10	1118.59	13	1178.2	5.059
R111	10	1096.72	14	1239.5	11.52
R112	9	982.14	13	1109.9	11.51

Table 39: Comparison of LSP Heuristic Solution with Best Heuristic Solution: C1XX.100 instances

Problem	Best Heuristic Solution		LSP Heuristic Solution		Solution
Instance	NV	Distance	NV	Distance	Gap %
C101	10	828.94	10	827.3	-0.2
C102	10	828.94	10	872.6	5.003
C103	10	828.06	11	898.4	7.829
C104	10	824.78	10	859.2	4.006
C105	10	828.94	10	827.3	-0.2
C106	10	828.94	10	850.4	2.524
C107	10	828.94	10	827.3	-0.2
C108	10	828.94	11	863.2	3.969
C109	10	828.94	10	860	3.612

2.9   Conclusions and Contributions

The primary contributions of this chapter include:

- Development of a fast, polynomial-time heuristic denoted the Layered Shortest Path (LSP) heuristic that generates near-optimal solutions to the elementary shortest path problem with resource constraints (ESPPRC);

- An analysis of the LSP heuristic, including its computational complexity, its utility as an optimal algorithm for the unconstrained shortest path problem (SPP), and an example demonstrating its non-optimality for resource-constrained versions of the

Table 40: Comparison of LSP Heuristic Solution with Best Heuristic Solution: RC1XX.100 instances

Problem Instance	Best Heuristic Solution		LSP Heuristic Solution		Solution Gap %
	NV	Distance	NV	Distance	
RC101	14	1696.94	16	1667.8	-1.75
RC102	12	1554.75	15	1555.1	0.023
RC103	11	1261.67	13	1411.2	10.6
RC104	10	1135.48	12	1294.1	12.26
RC105	13	1629.44	17	1641	0.704
RC106	11	1424.73	14	1479.2	3.682
RC107	11	1230.48	13	1359.2	9.47
RC108	10	1139.82	13	1310.2	13

Table 41: Comparison of LSP Heuristic Solution with Best Heuristic Solution: R2XX.100 instances

Problem Instance	Best Heuristic Solution		LSP Heuristic Solution		Solution Gap %
	NV	Distance	NV	Distance	
R201	4	1252.37	11	1222.5	-2.44
R202	3	1191.7	9	1078.2	-10.5
R203	3	939.54	8	1049.3	10.46
R204	2	825.52	7	908.5	9.134
R205	3	994.42	7	1105.9	10.08
R206	3	906.14	8	1060.5	14.56
R207	2	893.33	8	962.5	7.186
R208	2	726.75	6	991.3	26.69
R209	3	909.16	7	994.8	8.609
R210	3	939.34	8	1004.5	6.487
R211	2	892.71	8	920.5	3.019

Table 42: Comparison of LSP Heuristic Solution with Best Heuristic Solution: C2XX.100 instances

Problem Instance	Best Heuristic Solution		LSP Heuristic Solution		Solution Gap %
	NV	Distance	NV	Distance	
C201	3	591.56	3	589.1	-0.42
C202	3	591.56	4	634.9	6.826
C203	3	591.17	5	786.2	24.81
C204	3	590.6	5	815.8	27.6
C205	3	588.88	3	625.5	5.855
C206	3	588.49	5	648.6	9.268
C207	3	588.29	5	699.8	15.93
C208	3	588.32	4	661.8	11.1

Table 43: Comparison of LSP Heuristic Solution with Best Heuristic Solution: RC2XX.100 instances

Problem	Best Heuristic Solution		LSP Heuristic Solution		Solution
Instance	NV	Distance	NV	Distance	Gap %
RC201	4	1406.91	8	1290.9	-8.99
RC202	3	1367.09	8	1158	-18.1
RC203	3	1049.62	8	1115.8	5.931
RC204	3	798.41	8	1016	21.42
RC205	4	1297.19	11	1282.6	-1.14
RC206	3	1146.32	8	1149.5	0.277
RC207	3	1061.14	8	1090.9	2.728
RC208	3	828.14	6	904.7	8.462

problem;

- A computational study investigating the quality of solutions and the computational efficiency of using a standard root node column generation heuristic for solving the VRPTW, with the LSP heuristic used to solve the pricing subproblem. Results indicate that the LSP approach is a computationally attractive method for problems of practical size that yields good to very good solution quality, especially when the number of customers per tour is small.

CHAPTER III

UNCONGESTED PORT DRAYAGE

In this chapter, we consider a basic drayage routing and scheduling problem that will serve as the underlying problem that we will generalize in subsequent chapters. We provide a formal mathematical statement for the problem, and review related literature. We then analyze two versions of the problem, one with the objective of minimizing the number of vehicles required to complete a set of drayage tasks and another with the objective of minimizing the total travel time. We show that both problem variants are NP-hard combinatorial optimization problems. Next, we present a mixed integer programming formulation and discuss the complexities involved in finding optimal solutions for large practical instances. We then develop techniques for finding good solutions with reasonable computation times using a set covering model and a root column generation heuristic, where the pricing problem is solved with the LSP heuristic developed in Chapter 2. Finally, we conclude with computational results to show the quality and efficiency of the heuristic solution techniques developed for the drayage problem.

3.1 *Problem Definition*

We first present a formal mathematical definition of the uncongested drayage problem. Consider a drayage firm that must serve on a single day n_e export container move requests and n_i import requests. Let $\mathcal{C}$ be the set of container move requests, where $\mathcal{C}^E \subseteq \mathcal{C}$ is the set of export moves and $\mathcal{C}^I \subseteq \mathcal{C}$ is the set of import moves. Let $\mathcal{E} = \{E_1, E_2 \ldots, E_{n_e}\}$ represent the set of export customer locations and $\mathcal{I} = \{I_1, I_2 \ldots, I_{n_i}\}$ represent the set of import customer locations. Note that these locations are not necessarily unique. Each export move request $j \in \mathcal{C}^E$ represents a truck move from an origin customer location E_j to a port location P. Similarly, each import move request $k \in \mathcal{C}^I$ represents a move from P to a destination customer location I_k. Note that if a customer requests multiple container

47

moves, we will create a separate move request for each container.

The drayage company operates a fleet of vehicles based at a single depot location D, and develops routes and schedules for its vehicles given the set of move requests. Suppose first that the company uses a homogeneous fleet, and the number of vehicles in the fleet is not bounded. Each vehicle departs the depot, serves a sequence of requests, and then returns to the depot. Each request may have time window constraints at both the origin and destination of the move, and the vehicle serving the request must arrive at the appropriate node within the time window. The time window at the origin of request i is given by $[a_i^O, b_i^O]$ and the time window at the destination by $[a_i^D, b_i^D]$. In the models to follow, we allow vehicles to arrive at locations earlier than the start of a time window; in this case, they must dwell before beginning service. Note that when the duration of a request is fixed, compatible origin and destination time windows can be translated into a single time window $[a_i, b_i]$ at the destination of the request. Finally, the depot location D is also time-constrained; we assume that no vehicle may depart D before time 0, and that all vehicles must arrive back at D no later than the end of the operating period, time τ.

The depot, port, and each export and import location are modeled as points in a bounded region within a two-dimensional plane. Let t_{ij} represent the travel time between any two locations i and j. Travel times are symmetric: $t_{ij} = t_{ji}$.

We will denote this generic uncongested drayage problem setting by UDP. Given this setting, the drayage company must develop feasible routes for its vehicles in order to meet the objective of minimizing the transportation cost. In this dissertation, we investigate two different objective functions for the problem framework UDP. In the first case, we use minimizing the fleet size as the objective, and we denote this problem as $UDVP$. In the second case, we use minimizing the total travel time of all vehicles as the objective function, and denote this problem $UDTP$.

3.2   Background Literature

Problems $UDVP$ and $UDTP$ are specific variants of the class of problems denoted pickup and delivery problems with time windows (PDPTW). Dumas et al. (1991) defines a PDPTW as a problem for determining optimal routes from multiple depot locations to satisfy transportation requests, each requiring pickup at some origin and delivery at some destination, given constraints enforcing vehicle capacity, node time windows and precedence relationships. Each route also satisfies pairing constraints that guarantee that corresponding pickup and delivery locations are serviced by a route from same depot. Note that the generic vehicle routing with time windows problem (VRPTW) is a special case of PDPTW where the destination of each request is the unique depot location. Dumas et al. (1991) presents an optimal algorithm for solution of PDPTW problems using a column generation scheme with a shortest path subproblem with capacity, time window, precedence and coupling constraints. We refer the reader to Savelsbergh and Sol (1995) and Mitrovic-Minic (1998) for reviews of solution methods for various pickup and delivery problems. In general, multiple-vehicle pickup and delivery problems with time windows have received limited research attention.

The drayage problem considered in this research belongs to the subclass of PDPTW problems that require full truckloads; in such problems, each vehicle can only carry the load of a single customer at any given time. Gronalt et al. (2003) considers the problem of delivering full truckloads between different distribution centers operated by a logistics service provider, where each distribution center requires pickups as well as drop-offs of goods. This paper presents an integer programming formulation for the problem of minimizing total vehicle travel cost, and develops heuristic algorithms to generate good solutions for the problem. Lower bounds are provided by solving a linear programming formulation.

Port drayage routing is a restricted case of the PDPTW where each vehicle can transport a single load request at a time. In cases where each load request may be modeled as a single node, the problem reduces to an asymmetric multiple traveling salesman problem with time window constraints (m-TSPTW). These problems are frequently solved via Dantzig-Wolfe decomposition and Lagrangian relaxation techniques. Each of these techniques requires the solution of a time-constrained shortest path subproblem, which is also NP-hard (Dror, 1994).

Alternative approaches are explored in Wang and Regan (2002). This reference develops a solution method for an m-TSPTW for local truckload pickup and delivery in which the number of tasks assigned to each vehicle at any time is relatively small. An iterative solution procedure is developed using discretized time windows. The solution technique solves over-constrained and under-constrained versions of the problem to develop upper bound and lower bounds that are used in an iterative procedure. Tests focusing on small but operationally-realistic instances suggest that problems of reasonable size can be solved quickly.

3.3 Complexity of Problems UDVP and UDTP

In this section, we prove that both uncongested drayage problems defined in Section 3.1 belong to the class NP-hard. Note that the hardness of these problems is not guaranteed by the fact that they are special cases of the known hard problems discussed in Section 3.2. We prove hardness by showing that decision versions of each problem belong to the class NP-complete.

3.3.1 Complexity of Problem UDVP

To prove that the uncongested drayage problem with the objective of minimizing the number of vehicles required to complete the set of dray tasks (UDVP) is NP-hard, we first prove that the UDVP is in NP. We then select the Bin Packing Problem(BPP), which is an established NP-hard problem in the strong sense, and model a polynomial transformation from BPP to UDVP-R , a special case of UDVP . Problems of type UDVP-R represent drayage problems with only import requests, no task time windows, and the depot and port co-located.

We first define decision versions of both the Bin Packing Problem (BPP) and the restricted drayage problem UDVP-R.

Bin Packing Problem (BPP)

Instance: n objects with integer weights $d_1, d_2, \ldots, d_n$ to be placed in bins, each with integer capacity τ.

Decision Question: Is it possible to pack all n objects in $\leq m$ bins?

Uncongested Drayage Restricted Problem (UDVP-R)

Instance: an instance of UDVP with n import locations $I_1, I_2, \ldots, I_n$, each requesting a single container move from the port; port and depot co-located, *i.e.*, $P = D$; travel time t_{D,I_i} between $D = P$ and import node I_i is $\frac{d_i}{2}$; and a operating period duration given by τ.

Decision Question: Is it possible to feasibly complete all n import tasks with $\leq m$ vehicles?

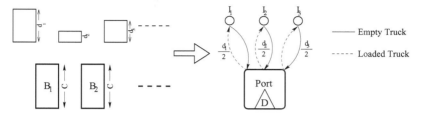

Figure 5: Transformation of BPP to UDVP-R

Note that given any instance of BPP, we can construct a related instance of UDVP-R in a polynomial number of steps. Figure 5 graphically depicts the transformation. We now show the equivalence of the decision problems UDVP-R and BPP.

Lemma 3.3.1. *If there exists a feasible packing for the BPP, then there is exists a feasible solution for UDVP-R.*

Proof. Consider any 'yes' instance of BPP with $k \leq m$ bins. Let N_j represent the items assigned to bin j in the solution. By definition of BPP, we have that:

$$\sum_{i \in N_j} d_i \leq \tau \tag{4}$$

By construction of the UDVP-R instance, it is clear that each import container move to location I_i requires d_i time units of a vehicle, since the vehicle must depart the port location, travel to I_i, then return to the port to retrieve the next container or to finish its daily operations at the depot. Each vehicle has τ time units available for daily operations. Hence from equation 4, we can observe that a solution to the UDVP-R in which N_j are the set of tasks assigned to vehicles $j = 1, \ldots, k \leq m$ is always feasible.

Thus a 'yes' instance of the BPP corresponds directly to a 'yes' instance for the UDVP-R. □

Lemma 3.3.2. *If there is exists a feasible solution to UDVP-R, then there exists a feasible packing for the BPP.*

Proof. Consider any 'yes' instance of UDVP-R using $k \leq m$ vehicles. Let N_j be the set of dray tasks performed by vehicle $j = 1, ..., k \leq m$ in the solution. By construction of the UDVP-R instance, we know that the total duration of each vehicle's assigned operations is less than τ:

$$\sum_{i \in N_j} d_i \leq \tau \tag{5}$$

¿From equation 5, we can observe that a solution to the BPP in which N_j are the set of items assigned to bins $j = 1, \ldots, k \leq m$ is always feasible.

Thus a 'yes' instance of the UDVP-R corresponds directly to a 'yes' instance for the BPP. □

Theorem 3.3.3. *UDVP is NP-hard.*

Proof. It is easy to see that the general UDVP is in NP, since verification of whether or not a solution is feasible can be performed in linear time in the number of move requests.

Lemmas 3.3.1 and 3.3.2 show that a yes instance of the decision version of the BPP can be obtained if and only if there exists a corresponding yes instance of UDVP-R. Thus, if it is possible to find a polynomial algorithm for the drayage problem UDVP, it is possible to solve the BPP in polynomial time. Since BPP is NP-complete, UDVP is NP-complete. Therefore, the optimization version of UDVP is NP-Hard. □

3.3.2 Complexity of Problem UDTP

To prove that the uncongested drayage problem with the objective of minimizing the total travel time required to complete the set of dray tasks (UDTP) is NP-hard, we first prove that the UDTP is in NP. We then again select the Bin Packing Problem(BPP) and model a polynomial transformation from BPP to UDTP-R, a special case of UDTP. Problems of type UDTP-R represent drayage problems with only import requests and no task time windows.

The decision version of BPP is given in Section 3.3.1. We now define a decision version of the restricted drayage problem UDTP-R.

Uncongested Drayage Restricted Problem (UDTP-R)

Instance - an instance of UDTP with n import locations $I_1, I_2, \ldots, I_n$, each with single container move request from the port; travel time t_{P,I_i} between the port P and import node I_i given as $\frac{d_i}{2}$ time units, travel time t_{DP} between the depot D and the port P given as $\frac{\epsilon}{2}$ time units (where $\epsilon > 0$), and travel time $t_{I_i,D}$ between import node I_i and the depot D given as $\frac{d_i + \epsilon}{2}$ time units; and an operating period duration given by $\tau + \epsilon$.

Decision Question - Is it possible to feasibly complete all n import tasks with total duration $T \leq m\epsilon + \sum_{i \in N} d_i$?

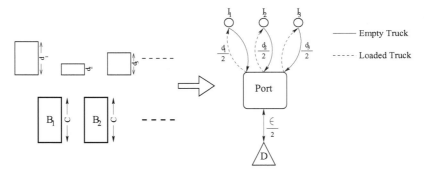

Figure 6: Transformation of BPP to UDTP-R

Note that given any instance of BPP, we can construct a related instance of UDTP-R in a polynomial number of steps. Figure 6 graphically depicts the transformation. We now show the equivalence of the decision problems UDTP-R and BPP.

Lemma 3.3.4. *If there exists a feasible packing for the BPP, then there is exists a feasible solution for UDTP-R.*

Proof. Consider any 'yes' instance of BPP with $k \leq m$ bins. Let N_j represent the items

53

assigned to bin j in the solution. By definition of BPP, we have that:

$$\sum_{i \in N_j} d_i \leq \tau \tag{6}$$

which of course implies that:

$$\epsilon + \sum_{i \in N_j} d_i \leq \epsilon + \tau \tag{7}$$

Now consider the related instance UDTP-R, and suppose we assign the import move requests N_j to each vehicle j. Each vehicle j departs the depot first for the port, requiring $\frac{\epsilon}{2}$ time units. Then, the vehicle serves each move request, traveling from the port to the importer location and back again for each request; for each import location I_i, such a cycle requires d_i time units. Finally, the vehicle returns from the port to the depot, requiring $\frac{\epsilon}{2}$ time units. Thus, the total time used by each vehicle j is given by:

$$t_j = \epsilon + \sum_{i \in N_j} d_i \tag{8}$$

and by (7), we see that each vehicle route is feasible.

Summing (8) over all k vehicles yields:

$$T_k = \sum_{j=1}^{k} t_j = k\epsilon + \sum_{i \in N} d_i \tag{9}$$

and since $k \leq m$,

$$T_k \leq T. \tag{10}$$

Thus a 'yes' instance of the BPP corresponds directly to a 'yes' instance for the UDTP-R. $\square$

Lemma 3.3.5. *If there is exists a feasible solution to UDTP-R, then there exists a feasible packing for the BPP.*

Proof. Consider any 'yes' instance of UDTP-R using $T_k \leq T$ time units, and suppose the solution uses k vehicles. Let N_j be the set of dray tasks performed by vehicle $j = 1, ..., k$ in the solution.

In problems UDTP-R, the lowest travel time for vehicle j serving import move requests N_j is to depart the depot for the port, then to serve each task in sequence with a round-trip from the port, and then finally to return to the depot. Note that a direct return to the depot from the final import customer location can never reduce the cost, by definition. Therefore, the travel time for vehicle j is given by:

$$\epsilon + \sum_{i \in N_j} d_i \leq \epsilon + \tau \qquad (11)$$

where the inequality in (11) holds by definition of UDTP-R.

Summing (11) for each vehicle yields:

$$k\epsilon + \sum_{i \in N} d_i \leq m\epsilon + \sum_{i \in N} d_i \qquad (12)$$

where the inequality holds since this is a 'yes' instance for UDTP-R.

We can construct a 'yes' instance of bin packing given this 'yes' instance for UDTP-R by using k bins, where bin j is assigned the items N_j. By (11), each bin j is feasible, and by (12) we have $k \leq m$.

Thus a 'yes' instance of the UDTP-R corresponds directly to a 'yes' instance for the BPP. □

Theorem 3.3.6. *UDTP is NP-hard.*

Proof. It is easy to see that the general UDTP is in NP, since verification of whether or not a solution is feasible can be performed in linear time in the number of move requests.

Lemmas 3.3.4 and 3.3.5 show that a yes instance of the decision version of the BPP can be obtained if and only if there exists a corresponding yes instance of UDTP-R. Thus, if it is possible to find a polynomial algorithm for the drayage problem UDTP, it is possible to solve the BPP in polynomial time. Since BPP is NP-complete, UDTP is NP-complete. Therefore, the optimization version of UDTP is NP-Hard. □

3.4 Mixed Integer Programming Formulations for UDVP and UDTP

In this section, mixed-integer programming formulations for problems under the framework UDP are presented. Separate formulations for UDVP and UDTP require only a minor change to the objective function.

For convenience, each container request is represented by a separate node located at the port location, P_i^E for export container request i and P_j^I for import request j. We now define a network model $G = (\mathcal{N}, \mathcal{A})$ for the problem. The node set $\mathcal{N}$ includes:

Depot Node: D

Exporter Location Nodes: $\mathcal{E} = \{E_1, E_2 \ldots, E_{n_e}\}$

Importer Location Nodes: $\mathcal{I} = \{I_1, I_2 \ldots, I_{n_i}\}$

Port Nodes: $\mathcal{P} = \mathcal{P}^E \cup \mathcal{P}^I = \{P_1^E, P_2^E \ldots, P_{n_e}^E\} \cup \{P_1^I, P_2^I \ldots, P_{n_i}^I\}$

The set of all nodes, $\mathcal{N}$, is then given by

$$\mathcal{N} = \{D\} \cup \mathcal{E} \cup \mathcal{I} \cup \mathcal{P}$$

Truck travel is possible between any two nodes, however, we can logically restrict some possibilities since the truck can only transport one container at a time. Each feasible node to node connection will comprise the arc set $\mathcal{A}$; we now define this feasible set using notation $(i, j) \in \mathcal{N} \times \mathcal{N}$:

Depot-to-exporter arcs: $i = D, j \in \mathcal{E}$

Depot-to-port import arcs: $i = D, j \in \mathcal{P}^I$

Exporter-to-port export arcs: $i = E_k, j = P_k^E \quad \forall\, k = 1, ..., n_e$

Port export-to-port import arcs: $i \in \mathcal{P}^E, j \in \mathcal{P}^I$

Port export-to-exporter arcs: $i \in \mathcal{P}^E, j \in \mathcal{E}$

Port export-to-depot arcs: $i \in \mathcal{P}^E, j = D$

Port import-to-importer arcs: $i = P_k^I, j = I_k \quad \forall\, k = 1, ..., n_i$

Importer-to-exporter arcs: $i \in \mathcal{I}, j \in \mathcal{E}$

Importer-to-port import arcs: $i \in \mathcal{I}, j \in \mathcal{P}^I$

Importer-to-depot arcs: $i \in \mathcal{I}, j = D$

Further, let arc set $\mathcal{A}_C$ contain all exporter-to-port export and port import-to-importer arcs; these will be referred to as *compulsory* arcs since some vehicle must traverse them in each feasible solution. Note that we exclude circular arcs that define return travel from port export node P_k^E to exporter E_k, and from importer I_k to port import node P_k^I. For each arc a, let t_a be its non-negative travel time.

Routing decisions are specified by binary decision variables x_a for all $a \in \mathcal{A}$, where

$$x_a = \begin{cases} 1 & \text{if some vehicle traverses arc } a \\ 0 & \text{otherwise} \end{cases}$$

Scheduling decisions are specified by continuous time variables s_i for all $i \in \mathcal{N}$, which represent the time at which the servicing vehicle arrives at node $i \in \mathcal{N} \setminus \{D\}$. Variable s_D will represent the latest time any vehicle arrives back at the depot.

Consider now the following formulation, which is derived from a standard VRPTW model. For convenience, define $head(a)$ and $tail(a)$ as the head node and tail node respectively for each arc $a \in \mathcal{A}$. Further, let $\mathcal{I}_i$ be the set of arcs a such that $tail(a) = i$, and let $\mathcal{O}_i$ be the set of arcs a such that $head(a) = i$.

Formulation 1: Uncongested Drayage Problem UDP

$$\text{(UDTP) minimize} \quad \sum_{a \in \mathcal{A}} t_a x_a \tag{13}$$

or

$$\text{(UDVP) minimize} \quad \sum_{a \in \mathcal{O}_D} x_a \tag{14}$$

subject to:

$$\sum_{a \in \mathcal{I}_i} x_a - \sum_{a \in \mathcal{O}_i} x_a \;=\; 0 \qquad \forall \; i \in \mathcal{N} \tag{15}$$

$$\sum_{a \in \mathcal{O}_D} x_a \;\leq\; n_e + n_i \tag{16}$$

$$s_{head(a)} + t_a - M(1 - x_a) \;\leq\; s_{tail(a)} \quad \forall \; a \in \mathcal{A} \mid head(a) \neq D \tag{17}$$

$$t_a - M(1 - x_a) \;\leq\; s_{tail(a)} \quad \forall \; a \in \mathcal{A} \mid head(a) = D \tag{18}$$

$$a_i^O \leq \; s_i \; \leq \; b_i^O \qquad \forall \; i \in \mathcal{E} \cup \mathcal{P}^I \tag{19}$$

$$a_i^D \leq \; s_i \; \leq \; b_i^D \qquad \forall \; i \in \mathcal{I} \cup \mathcal{P}^E \tag{20}$$

$$x_a \;=\; 1 \qquad \forall \; a \in \mathcal{A}_C \tag{21}$$

$$s_D \;\leq\; \tau \tag{22}$$

$$x_a \;\in\; \{0,1\} \quad \forall \; a \in \mathcal{A} \tag{23}$$

$$s_i \;\geq\; 0 \qquad \forall \; i \in \mathcal{N} \tag{24}$$

Constraints (15) ensure vehicle flow balance at all nodes, while constraint (16) sets an upper bound for the number of vehicles used as the total number of container load requests. Constraints (17) and (18) use a big-M method to ensure consistency of vehicle arrival times at nodes. If and only if a vehicle traverses arc a, then the time the vehicle arrives at the tail node must be no earlier than the time it arrives at the head node plus the arc travel time. Additionally, for arcs terminating at the depot, constraints (17) act to set s_D to the latest arrival time back at the depot of any vehicle.

Constraints (19) ensure that the arrival times satisfy time window constraints. Constraints (21) is the compulsory arc constraint which guarantees that all containers are transported from their origin to their destination. Lastly, constraint (22) ensures that trucks return to the depot by τ.

This formulation can clearly be modified to model a linear combination of the cost functions of UDTP and UDVP, using the UDTP objective function. Arc travel time functions can be converted to costs through some multiplier; a fixed dispatch or vehicle cost can be then added to each arc originating from the depot. To purely minimize the total number of vehicles required as in problem UDVP, the travel time multiplier can be set to zero.

58

The complexity and the overwhelming size of practical drayage routing instances makes this formulation difficult to solve directly using commercial integer programming solvers. Hence we adopt a set covering approach to model the problem for practical instances with a large number of container move requests. The next section describes this model.

3.5 *Set Covering Models for UDVP and UDTP*

We develop set covering integer programming formulations for the problems UDVP and UDTP. Given the set $\mathcal{C}$ of all move requests ($n = |\mathcal{C}|$) and the set $\mathcal{R}$ of all feasible single-vehicle routes serving subsets of $\mathcal{C}$, a set covering model is formulated and solved to determine the minimum-cost subset of $\mathcal{R}$ which ensures that each customer request is served.

Let α_{ij} be a $\{0,1\}$ parameter equal to one if request i is served by route j, and let t_j be the total time required by route j. The decision variables x_j indicate which routes in $\mathcal{R}$ are chosen for the optimal subset:

$$x_j = \begin{cases} 1 & \text{if route } j \text{ is in final optimal solution} \\ 0 & \text{otherwise} \end{cases}$$

The set covering binary integer programming formulation for UDTP is then:

$$\text{minimize} \quad \sum_{j \in \mathcal{R}} t_j x_j$$

subject to:

$$\sum_{j \in \mathcal{R}} \alpha_{ij} x_j \geq 1 \quad \forall\ i \in \mathcal{C} \tag{25}$$

$$x_j \in \{0,1\} \quad \forall\ j \in \mathcal{R} \tag{26}$$

Simply replacing the objective function with $\sum_{j \in \mathcal{R}} x_j$ yields a formulation for UDVP.

3.6 *Solving Set Covering Models via Smart Enumeration*

We first present a method for effective *a priori* enumeration of all routes in the set $\mathcal{R}$ for problems within the framework UDP. The approach works well for small problems. This enumeration algorithm is a label generation approach that generates labels using a list propagation method. Each label is an ordered list of container move requests; here we

59

will use the customer location to represent the request. For example, a label might be $\{E_1, I_2, E_3\}$ which would represent a vehicle route that proceeds as follows: $D - E_1 - P - I_2 - E_3 - P - D$.

A label is taken from the beginning of the label list for potential extension, and all the new labels generated by extending this label are appended to the end of the label list. For example, label $\{E_1, I_2\}$ representing route $D - E_1 - P - I_2 - D$ may be feasibly extended to label $\{E_1, I_2, E_3\}$. The algorithm continues by removing the label being extended from the list and picking the next list element for extending, until no more labels are generated and the list is empty. Each label generated using this algorithm represents a feasible route, and thus the set of all labels represent the set of all feasible routes.

Standard enumeration techniques for NP-Hard routing problems with time window constraints consider routes serving the same set of customers in different sequences as different routes. For example, a normal enumeration algorithm will consider two routes given by C1-C2-C3 and C1-C3-C2 where C1, C2, and C3 represent customers as two separate routes.

The smart enumeration algorithm makes use of the problem structure of the UDP to find routes which dominate other routes. If multiple routes serve the same set of customers and finish at the same physical location, then the route with smaller completion time dominates other routes with higher current completion times. Note that a label is said to *finish* at the location prior to the depot. For example, the label $\{E_1, I_2, E_3\}$ finishes at the port location. The dominance search is made computationally efficient by limiting the search to a subset of routes.

Details of the smart enumeration algorithm are described below in the following subsection.

3.6.1 Algorithm

Let $\mathcal{L}$ denote the list of labels and n_L be the total number of labels generated by the algorithm. Each label $\ell \in \mathcal{L}$ consists of the following components:

- A route vector $\mathbf{r}^\ell$ with each element

$$r_i^\ell = \begin{cases} 1 & \text{if move request } i \in \mathcal{C} \text{ is covered by } \mathbf{r}^\ell \\ 0 & \text{otherwise} \end{cases}$$

- The last move request processed by $\mathbf{r}^\ell$, denoted by f^ℓ

- The time of completion of task f^ℓ, denoted by τ^ℓ

Let p^ℓ represent the total number of customers serviced by route $\mathbf{r}^\ell$ associated with label ℓ, $\mathcal{L}_p \subset \mathcal{L}$ be the set of labels serving p customers. Labels will be generated in such a way that all labels with p customers will be positioned in $\mathcal{L}$ before any label with $p+1$ customers. Let N_p be the label number of the first label with p customers. We now present the algorithm.

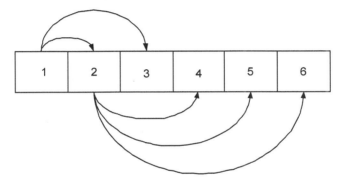

Figure 7: Label generation approach via extension.

Label Enumeration via Extension

Initialization:

1: Create initial label ℓ_0 at depot with no predecessor, with all attributes corresponding to ℓ_0 set to 0

2: Initiate $p = 0$, $n = 0$, and $N_0 = 0$

3: $LIST = \{\ell_0\}$

4: $\mathcal{L} = \{\}$

Iterations:

5: **while** $LIST \neq \emptyset$ **do**

6:    Remove first element ℓ from $LIST$

7:    **for all** $j \in \mathcal{C}$ not in route $\mathbf{r}^\ell$ **do**

8:      **if** j can be feasibly added to the route by departing the destination of request f_ℓ at time τ_ℓ **then**

9:        Generate new label ℓ' at j

10:        Append label ℓ' to the end of $LIST$ and $\mathcal{L}$

11:        $n \to n + 1$

12:        **if** $\sum r_i^{\ell'} = p + 1$ **then**

13:          $p \to p + 1$ and $N_{p+1} = n$

14:        **end if**

15:        RemoveDominatedLabels()

16:      **end if**

17:    **end for**

18: **end while**

Each new label ℓ' generated will be assigned attributes as follows:

- $f_{\ell'} = j$

- $r_i^{\ell'} = r_i^{\ell} \quad \forall \; i \in \mathcal{C} \setminus \{j\}$

- $r_j^{\ell'} = 1$

and

$$\tau_\ell' = \tau_\ell + \begin{cases} t_{Pj} + t_{jP} & \text{if } f_\ell \in \mathcal{E}, \; j \in \mathcal{E} \\ t_{Pj} & \text{if } f_\ell \in \mathcal{E}, \; j \in \mathcal{I} \\ t_{f_\ell j} + t_{jP} & \text{if } f_\ell \in \mathcal{I}, \; j \in \mathcal{E} \\ t_{f_\ell P} + t_{Pj} & \text{if } f_\ell \in \mathcal{I}, \; j \in \mathcal{I} \end{cases}.$$

3.6.1.1 Feasibility Conditions

The following conditions must be satisfied in order to successfully generate a new label ℓ' at j from label ℓ with current last task f_ℓ:

1. The new task j should not already be contained in route $\mathbf{r}^\ell$;

2. If $j \in \mathcal{C}^E$, then the vehicle serving route $\mathbf{r}^\ell$ should be able to travel from the destination of f_ℓ to E_j, departing at time τ_ℓ, and reach E_j within the permitted pickup time window $[a_j^O, b_j^O]$, then travel from E_j to P and arrive within the delivery time window $[a_j^D, b_j^D]$. Further, the vehicle must also be able to depart P and return to D by the deadline τ;

3. If $j \in \mathcal{C}^I$, then the vehicle serving route $\mathbf{r}^\ell$ should be able to travel from the destination of f_ℓ (which might be the port) to the port, reach P within the required pickup window $[a_j^O, b_j^O]$, then travel to I_j and arrive by the permitted delivery time window $[a_j^D, b_j^D]$. Further, the vehicle must also be able to depart I_j and arrive back at D by deadline time τ.

3.6.2   Routine to Remove Dominated Labels

The label generation procedure prevents creation of routes that are not necessary for determination of an optimal solution; such routes are represented by labels that cover the same tasks and terminate at the same physical location at different completion times. Consider, for example, a problem with 2 export tasks. Let label ℓ_1 correspond to route $D - E2 - E1$ and ℓ_2 to $D - E1 - E2$. If $\tau_{\ell_2} \geq \tau_{\ell_1}$, then ℓ_1 dominates ℓ_2 since both routes cover the same move tasks and terminate (currently) at the port.

In general, two routes $\mathbf{r}^\ell$ and $\mathbf{r}^{\ell'}$ finish at the same current physical location if

- Both routes have the same import task as their last tasks, $i.e.,$, $f_\ell = f_{\ell'}$, in which case they both currently terminate at the same import customer location; or

- Both routes have any of the export tasks as their last tasks, $i.e.,$, $f_\ell, f_{\ell'} \in \mathcal{C}^E$ in which case they both currently terminate at the port.

One of the properties of the label enumeration procedure is the non-decreasing value of p along the list, *i.e.*, , labels serving $p+1$ customers will be generated only after generating all possible labels serving p customers. This is due to the fact that new labels are always appended to the end of the list, and the label to be processed is taken from the beginning of the list. This property can also enable narrowing the search for dominated labels within each set $\mathcal{L}_p$, instead of searching through all the labels in $\mathcal{L}$.

Hence the dominance check can be performed in two steps:

1. Find a $\ell \in \mathcal{L}_p$ which has the identical route $\mathbf{r}^\ell$ to that of the new label $\mathbf{r}^{\ell'}$.

2. If both routes are currently terminate at the same physical location, check completion times for dominance.

The formal routine is presented below:

RemoveDominatedLabels()

1: Given new label ℓ' serving p tasks

2: **for all** Existing labels $\ell \in LIST(N_p, ..., n)$ **do**

3:     **if** $\mathbf{r}^{\ell'} = \mathbf{r}^\ell$ **then**

4:         **if** $f_{\ell'} = f_\ell$ or $f_{\ell'}, f_\ell$ both export tasks **then**

5:             **if** $\tau_{\ell'} \geq \tau_\ell$ **then**

6:                 Remove label ℓ' from $LIST$ and $\mathcal{L}$

7:             **else**

8:                 Remove label ℓ from $LIST$ and $\mathcal{L}$

9:                 $n \rightarrow n - 1$

10:             **end if**

11:         **end if**

12:     **end if**

13: **end for**

The removal of dominated labels helps to reduce the size of the resultant set covering model. This results in in reduced problem read time for commercial solvers, and faster convergence to the optimal solution using branch-and-bound algorithms.

Once a full set of labels are generated using this procedure, they can be added as the required column set $\mathcal{R}$ in the set covering formulation for UDTP or UDVP. For a given label ℓ, the tasks covered by the column are given by the route vector $\mathbf{r}^\ell$, and its cost in the case of UDTP is given by

$$t_\ell = \tau_\ell + \begin{cases} t_{PD} & \text{if } f_\ell \in \mathcal{E} \\ t_{f_\ell D} & \text{if } f_\ell \in \mathcal{I} \end{cases}$$

The enumeration technique performed well for smaller problem sizes, providing optimal results for problem sizes up to 30 customers. The detailed performance of this technique is presented in Section 3.8, where the results are used to assess the quality of heuristic solution procedures developed in this research.

3.7 Solving Set Covering Models with Root Column Generation Heuristic

We prove earlier that problems UDVP and UDTP are NP-hard optimization problems. The number of steps required for the complete enumeration of all feasible routes for this problem is an exponential function of the problem size, which makes it intractable to generate the set $\mathcal{R}$ for realistically-sized problems. Since $\mathcal{R}$ will contain a very large number of routes for such instances, we develop solution heuristics based on column generation, initially described in Dantzig and Wolfe (1960). In this work, we will specifically develop a root column generation heuristic. As explained in Chapter 2, such an approach uses heuristic column generation to solve approximately the linear relaxation of the set covering integer program. The initial set of routes, $\mathcal{R}'$, contains all routes that cover a single customer request; note that if any such routes are infeasible, then the associated customer request cannot be covered by any route. Once the heuristic column generation is complete, branch-and-bound is used to determine an integer solution to the problem using only the final set of columns $\mathcal{R}'$ from the column generation. Computational results in which we compare true optimal solutions to solutions generated via the root column generation heuristic for small problems indicate, however, such an approach can identify very good quality solutions.

In this research, we utilize the LSP heuristic developed in Chapter 2 to solve the pricing

65

subproblems associated with the set covering formulations for problems UDVP and UDTP. We now provide a description for how to apply the LSP heuristic to address these subproblems. The goal of the pricing subproblem is to identify columns with negative reduced cost to add to the current column subset considered by the master linear program. Consider for exposition problem UDTP. At some iteration, after solving the linear relaxation of P over $\mathcal{R}'$, let π_i represent the dual variables associated with constraints (25). Then, the reduced cost $\bar{c}_j$ of any route $j \in \mathcal{R}$ is given by

$$\bar{c}_j = t_j - \sum_{i \in \mathcal{C}} \alpha_{ij} \pi_i. \tag{27}$$

For problem UDVP, a similar reduced cost definition can be used where t_j in the above equation is replaced by the value 1.

Identifying columns with negative reduced cost can be accomplished by solving the following subproblem:

$$\min_{j \in \mathcal{R}} \bar{c}_j. \tag{28}$$

If the objective function is non-negative, no cost-improving column exists. Alternately, if the objective function is negative, then there exists at least one route j^* such that $\bar{c}_{j^*} < 0$ to be added to the column set $\mathcal{R}'$.

Problem (28) can be cast as an elementary shortest path problem with time-window constraints, which is a special case of the ESPPRC. The goal is to find an elementary path which begins at the vehicle depot D and ends at the vehicle depot D, serving a set of customers feasibly with minimum reduced cost[1].

3.7.1 Using the LSP Heuristic for Solution of UDP

We now describe how to use the LSP heuristic to develop suboptimal solutions to the pricing subproblems for UDTP and UDVP given by (28). In the exposition to follow, we only describe application of the heuristic to the problem UDTP; the procedure for UDVP is very similar, with only minor changes required to the arc cost definitions. Although the LSP

---

[1]Note that as written, such a path is not elementary since the depot node D is visited twice. This is a minor technical issue that can be easily resolved by splitting the depot into two separate nodes, one that originates the path and one that terminates the path.

heuristic is not guaranteed to return optimal solutions, it is a very efficient procedure, with worst-case complexity of $\Phi((n_e+n_i)^3)$ when applied to UDP problems; in such problems, the maximum number of layers will be n_e+n_i. Further, the method can potentially identify large numbers of routes with negative reduced cost each iteration; in the root column generation heuristic, all such routes are added to the current route set $\mathcal{R}'$ before the linear program is resolved.

To apply the LSP heuristic, we begin by defining the network $G = (\mathcal{V}, \mathcal{A})$ over which the method will determine paths. The set $\mathcal{V}$ will include a node for each container move request, as represented by the import or export customer location of the request, and a node for the container depot: $\mathcal{V} = \{D\} \cup \mathcal{E} \cup \mathcal{I}$. Arcs connecting all locations i and j in $\mathcal{V}$ are created to form $\mathcal{A}$. To solve UDP problems, we need only consider a single resource type, time. Therefore, for each arc (i, j) we need to define its time consumption $\bar{t}_{ij}$ in addition to its cost $\bar{c}_{ij}$:

$$
\bar{t}_{ij} =
\begin{cases}
t_{Dj} + t_{jP} & \text{if } i = D,\ j \in \mathcal{E} \\
t_{DP} + t_{Pj} & \text{if } i = D,\ j \in \mathcal{I} \\
t_{Pj} + t_{jP} & \text{if } i \in \mathcal{E},\ j \in \mathcal{E} \\
t_{Pj} & \text{if } i \in \mathcal{E},\ j \in \mathcal{I} \\
t_{ij} + t_{jP} & \text{if } i \in \mathcal{I},\ j \in \mathcal{E} \\
t_{iP} + t_{Pj} & \text{if } i \in \mathcal{I},\ j \in \mathcal{I} \\
t_{PD} & \text{if } i \in \mathcal{E},\ j = D \\
t_{iD} & \text{if } i \in \mathcal{I},\ j = D
\end{cases}
\tag{29}
$$

Given $\bar{t}_{ij}$, we can define the arc reduced costs for problem UDTP as follows:

$$
\bar{c}_{ij} =
\begin{cases}
\bar{t}_{ij} - \pi_j & \text{if } i \in \mathcal{V},\ j \in \mathcal{E} \cup \mathcal{I} \\
\bar{t}_{ij} & \text{otherwise}
\end{cases}
\tag{30}
$$

Alternatively, the arc reduced costs for UDVP are:

$$
\bar{c}_{ij} = \begin{cases} 1 - \pi_j & \text{if } i = D, \ j \in \mathcal{E} \cup \mathcal{I} \\ -\pi_j & \text{if } i \in \mathcal{E} \cup \mathcal{I}, \ j \in \mathcal{E} \cup \mathcal{I} \\ 0 & \text{otherwise} \end{cases} \tag{31}
$$

For UDP problems, the start node for path generation is the vehicle depot: $v_0 = D$. Again, each label ℓ_{kj} will be used to store information about the layered shortest path $P^*_{v_0 j}(k)$. One additional label $\ell_{0 v_0}$ is used for initialization. Let $\mathcal{L}$ denote the set of all labels ℓ_{kj} generated by the heuristic, and $\mathcal{L}^A$ be the set of labels corresponding to negative reduced cost paths. Each label $\ell_{kj} \in \mathcal{L}$ contains the following attributes:

- A path vector $\mathbf{p}_{kj}$ of length $n_e + n_i$ with each element

$$
p^i_{kj} = \begin{cases} 1 & \text{if customer request } v_i \in \mathcal{E} \cup \mathcal{I} \text{ is already covered in path } P^*_{v_0 j}(k) \\ 0 & \text{otherwise} \end{cases} ;
$$

- The current arrival time of path $P^*_{v_0 j}(k)$ at the destination node of task j, denoted by τ_{kj} ; and

- The cost of path $P^*_{v_0 j}(k)$, given by δ_{kj}.

LSP Heuristic

Initialization:

1: Initialize label ℓ_{0,v_0} corresponding to the start node v_0 representing an empty path, set all attributes of ℓ_{0,v_0} to 0

2: $\mathcal{L}^A \rightarrow \emptyset$

   Iterations:

3: **for all** $j \in \mathcal{V} \setminus \{v_0\}$ **do**

4:    **if** $checkExtend(\ell_{0,v_0}, j) = $ TRUE **then**

5:       $layerExtend(\ell_{0,v_0}, j)$

6:    **end if**

7: **end for**

8: $k = 2$

9: **while** $k \leq |\mathcal{V}| - 1$ **do**

10:    **for all** $i \in \mathcal{V} \setminus \{v_0\}$ such that $\ell_{k-1,i}$ exists **do**

11:      **for all** $j \in \mathcal{V} \setminus \{v_0\}$ **do**

12:        **if** $checkExtend(\ell_{k-1,i}, j) = \text{TRUE}$ **then**

13:          $layerExtend(\ell_{k-1,i}, j)$

14:        **end if**

15:      **end for**

16:    **end for**

17:    $k = k + 1$

18: **end while**

$checkExtend(\ell_{ki}, j)$

1: **if** $p_{ki}^{j} = 0$ **then**

2:    **if** $\tau_{ki} + \bar{t}_{ij} \leq b_j$ **then**

3:      **if** $\max(a_j, \tau_{ki} + \bar{t}_{ij}) + \bar{t}_{jD} \leq \tau$ **then**

4:        **if** $\ell_{k+1,j}$ does not exist **then**

5:          Generate blank label $\ell_{k+1,j}$

6:          return TRUE

7:        **else**

8:          **if** $\delta_{ki} + \bar{c}_{ij} \leq \delta_{k+1,j}$ **then**

9:            return TRUE

10:          **end if**

11:        **end if**

12:      **end if**

13:    **end if**

14: **end if**

15: return FALSE

$layerExtend(\ell_{ki}, j)$

1: $\mathbf{p}_{k+1,j} = \mathbf{p}_{ki}$

2: $p_{k+1,j}^{j} = 1$

3: $\delta_{k+1,j} = \delta_{ki} + \bar{c}_{ij}$

4: $\tau_{k+1,j} = max(a_j, \tau_{ki} + \bar{t}_{ij})$

5: **if** $\delta_{k+1,j} + \bar{c}_{jD} < 0$ **then**

6: $\quad \mathcal{L}^A \rightarrow \mathcal{L}^A \cup \{\ell_{k+1,j}\}$

7: **end if**

At the conclusion of the heuristic, we add all labels in the set $\mathcal{L}^A$ corresponding to routes with negative reduced costs to $\mathcal{R}'$. If $\mathcal{L}^A = \emptyset$, we have found no improving routes and thus the heuristic column generation is complete.

3.8 Computational Results

In this section, we present a set of computational results that indicate that the approaches developed are effective. Solution techniques for the uncongested drayage problem UDTP were tested on random problem instances representative of typical container port operations. The solution methods were implemented in the C programming language, and utilize the CPLEX Version 8.0 callable libraries for the solution of linear and binary integer programs when necessary. All tests were run on a dual-CPU 2.4 GHz Pentium with 2 GB of memory running Linux. Computation times in the tables to follow are given in seconds.

The heuristics are able to solve problems with up to 100 transportation requests within 7-8 minutes of CPU time on typical PC hardware. The methods, therefore, appear to be practical for typical problem sizes faced in the industry. The quality of the solutions returned by the root column generation heuristic is evaluated by comparison with true optimal solutions obtained, with much greater computational effort, using the enumerative approach for problems with fewer customer requests. When using the fast LSP pricing heuristic, the root column generation heuristic approach generates results with optimality gaps of no greater than 1.5% for a representative set of 8 different 30-customer problems.

3.8.1 Data Generation

To test our methods on the hardest drayage problems, we assume that all container move requests face a common task completion deadline of t_d; thus, the time window $[a_i, b_i] = [0, t_d]$ for each request i. Although the instances therefore do not represent the full range of potential drayage problems, they do model a common variant in which containers must be delivered to port to meet a ship sailing deadline. Further, since all move requests face a common deadline, there are many feasible routes and therefore these problems are in some sense the most difficult that need to be solved in practice.

For each problem size, 10 different instances were tested. To generate an instance, the location coordinates of each exporter and importer warehouse are randomly generated using a random two-dimensional uniform distribution generating function specified over a rectangular region. The location of the port and the depot are fixed at two points inside the rectangular region for all problem data sets. Travel times are specified by an L_1 distance function.

The following parameters are used for all generated problem instances:

Rectangular service region: $x \in [0, 2], \ y \in [0, 1]$

Depot location: $(1, \frac{1}{2})$

Port location: $(1, 0)$

Task completion deadline $= t_d = 6$ hours

Vehicle depot return deadline $= \tau = 9$ hours

Note that the dimensions of the service region are measured in hours of travel time.

3.8.2 Performance of Smart Enumeration

Table 44 demonstrates the impact of the dominance check on the enumeration algorithm. The results in the table represent averages over 10 different random instances with 20 customers each (10 export requests and 10 import requests). By using the dominance check, the total number of enumerated routes was reduced by more than 90%. Note that

71

for this set of problems, however, the average total CPU time remained roughly the same with and without the dominance check. Although the time required by CPLEX to solve the integer programs generally decreases when using the dominance check to remove unnecessary columns, performing the check requires more route generation time.

The primary benefit of the dominance check is that it enables the enumerative approach to handle larger problem sizes. The reduction in the number of variables allowed CPLEX to handle set covering models for larger problem sizes, and optimal solutions were obtained for problem instances with up to 30 customers.

Table 44: Performance of Smart Enumeration Solution Approach for UDTP: With and Without Dominance Check

Data Set No.	Without Dominance Check						With Dominance Check					
	Labels Generated	Generation Time	CPlex Time			Total Time	Labels Generated	Generation Time	CPlex Time			Total Time
			Read	Presolve	B&B				Read	Presolve	B&B	
1	420415	1.91	5.81	17.67	221.22	246.61	36800	18.35	0.26	0.89	160.04	179.54
2	1945055	10.32	69.83	94.04	26.39	200.58	53980	40.6	0.4	1.54	20.72	63.26
3	445770	2.18	6.63	19.37	28.79	56.97	40706	19.87	0.32	1.04	30.51	51.74
4	1245089	6.19	32.78	61.27	510.09	610.33	64715	45.46	0.47	1.94	674.92	722.79
5	65412	0.35	0.54	2.9	30.76	34.55	16070	5.3	0.06	0.25	72.05	77.66
6	1288118	6.36	34.04	63.5	200.56	304.46	80964	106.62	0.64	2.37	117.41	227.04
7	1279167	6.54	33.15	59.39	11127	11226	65107	49.76	0.48	2.01	N.A	N.A
8	277199	1.3	3.48	12.37	1.78	18.93	30401	14.67	0.19	0.72	1.19	16.77
9	175182	0.84	1.6	7.29	0.38	10.11	29556	10.69	0.16	0.65	0.26	11.76
10	251091	1.13	2.82	11	6.2	21.15	32645	14.71	0.21	0.75	7.76	23.43
Average*	679259	3.4	17.5	32.2	114	167.1	42870.8	30.7	0.3	1.1	120.6	152.7

* Average calculation does not include instance 7

73

Note that instance number 7 with the dominance check did not run to completion during the branch-and-bound.

3.8.3 Performance of Heuristics

The performance of the column generation heuristics is now evaluated by comparing the solutions obtained from the heuristic with the optimal solutions obtained from the smart enumeration method. Table 45 summarizes the performance of LSP pricing heuristic on UDTP problems. Table 46 compares the computational time required by the two methods for the same problem sizes; note that computation times are decomposed into label generation time and branch-and-bound time. All values represent averages over 10 instances.

Table 45: Performance of Root Column Generation Heuristic Using LSP for UDTP: Solution Quality

Problem Size	No. of Instances solved optimally	Avg optimality gap	Avg Requests per vehicle
11 X 11	5	0.586%	4.72
12 X 12	1	1.921%	4.72
13 X 13	4	0.757%	4.42
14 X 14	3	1.252%	4.47
15 X 15	1	1.451%	4.82

Table 46: Performance of Root Column Generation Heuristic Using LSP for UDTP: Computation Time

Problem Size	Smart Enumeration		LSP Heuristic	
	Label Generation	B & B	Label Generation	B & B
11 X 11	27.5	2925.0	0.12	0.20
12 X 12	87.9	2574.5	0.17	0.57
13 X 13	95.6	22813.3	0.22	1.37
14 X 14	306.1	17419.6	0.33	1.93
15 X 15	803.8	8817.2	0.56	17.50

Table 45 clearly demonstrates that the column generation heuristic is able to solve small problems to near-optimality; the reported average optimality gap is with respect to the known optimal solutions to these problems found using the smart enumeration method. Further, Table 46 demonstrates that the heuristic solution approach is much faster than the exact enumerative method.

Table 47 summarizes the performance of the LSP heuristic for larger problem sizes. Importantly, the method is able to generate solutions to problems with 100 move requests in under 5 minutes. Further, since the average numbers of requests served per vehicle are consistent with those for smaller problems, we suspect that the solution quality remains high.

Table 47: Performance of Root Column Generation Heuristic Using LSP for UDTP: Large Problems

| Problem Size | CPU Time | | Avg Requests per vehicle |
	Label Generation	B & B	
20 X 20	2.3	4.0	4.46
30 X 30	20.3	33.2	4.76
40 X 40	87.4	91.9	4.75
50 X 50	289.3	91.7	4.77

Table 48 shows the improvement in solutions obtained by using the k-LSP pricing heuristic. Many problems can be solved to optimality with this heuristic using small values of k, however some problems remain difficult.

Table 48: Comparison of Pricing Heuristics for 30-request Problems

Data Set Number	Smart Enumeration			LSP Heuristic			k-LSP Heuristic			
	Solution Value	Label Gen. Time	Total Time	Solution Value	Label Gen. Time	Total Time	Solution Value	Label Gen. Time	Total Time	k Value
1	37.256	505.4	527.9	38.282	0.54	1.29	37.418	1345.4	1345.5	150
2	36.618	1783.1	19767.1	36.912	0.70	0.88	36.618	77.0	77.5	26
3	36.048	1003.7	1027.2	36.294	0.67	0.73	36.048	2.0	2.0	6
4	39.312	279.7	26569.1	39.334	0.48	0.61	39.312	28.7	75.7	20
5	36.512	1758.0	1868.8	37.522	0.50	1.37	36.742	799.7	799.9	150
6	40.114	202.0	26268.8	40.114	0.51	1.63	40.114	0.5	1.7	1
7	37.498	873.8	910.7	38.404	0.61	17.16	37.764	1169.0	1169.2	150
8	45.064	24.9	28.2	46.024	0.43	120.80	45.064	5.5	5.6	11

3.9 Conclusions and Contributions

The primary conclusions and contributions of this chapter include:

- A review of the literature focusing on pickup and delivery problems with full truckloads under time window constraints, a generalization of the basic problem considered in this research;

- Development of decision support models a basic drayage routing and scheduling problem;

- Construction of proofs that uncongested drayage routing and scheduling problems, which are special cases of several well-known hard problems, are in the class NP-hard. These results motivate the search for effective heuristic solution procedures;

- Development of both exact and heuristic solution approaches for large, realistically-sized drayage routing and scheduling problems based on a set covering formulation and column generation;

- Computational study verifying the quality and computational efficiency of the heuristic column generation solution approaches. The study shows that near-optimal solutions to port drayage pickup and delivery problems can be generated with reasonable computation times.

CHAPTER IV

PORT DRAYAGE WITH TIME-DEPENDENT ACCESS DELAY

In this chapter, we discuss an extension of the drayage problem studied in Chapter 3 that we denote the congested drayage problem (CDP). In the CDP, we assume that access to the port is congested, and that each drayage port access experiences some non-negative delay. In this research, we limit analysis to the case where port access delay is known, given the time of day of the port access. We begin the chapter with a formal problem statement, and then survey existing literature for similar variants of generic routing problems. Next, the mixed integer linear programming model developed in 3.4 for the UDP is extended to enable modeling of CDP problems with piecewise-linear delay functions. To handle problems of realistic size, we then extend the heuristic techniques developed in Chapter 2 for application to the CDP. We conclude the chapter with a set of computational results on the impact of access congestion on drayage operations.

4.1 *Problem Definition*

Truck traffic into and out of seaports tends to fluctuate over the course of a day. As might be expected, traffic peaks often occur before departing vessel *cut times*, which are deadlines for departing outbound containers. Other time-of-day phenomena also create traffic peaks. When drayage drivers attempt to access ports during congested time periods, they experience delay. Since such delays increase the time needed to complete tasks, drayage companies require additional driver/tractor resources to serve similar sets of tasks when operating at congested ports.

We extend the definition of the problem framework UDP defined in Section 3.1 to incorporate delays caused by congested access to the port. To model such delays, each vehicle (empty or loaded) arriving at the entry gate to the port is assumed to be delayed according to a time-dependent, exogenously-defined congestion function. We assume that this function is known with complete certainty when planning; for example, the company may have access to historical time-of-day congestion delay statistics.

In this research, we utilize a congestion function that maps the arrival time t of a vehicle at the port node to a time delay value $f(t)$. We assume that time windows (pickup for import container requests, and dropoff for export container requests) are enforced after the delay is added; the effective arrival time of a vehicle at the port is therefore $t + f(t)$. Given this setting, the generic CDP problem for the drayage company is to develop feasible routes for its vehicles to meet the objective of minimizing transportation cost.

4.1.1 FIFO Property of Delay Function

Similar to Ahn and Shin (1991), we assume that any feasible congestion function will satisfy a first-in first-out (FIFO) property. Essentially, such a property will guarantee that if vehicle j arrives at the congested port node after vehicle i, then vehicle j can depart no earlier than i.

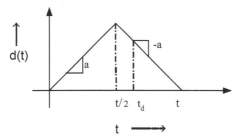

Figure 8: A congestion function $f(t)$ satisfying the FIFO property

We can represent this condition mathematically as follows:

Definition 5 (FIFO Congestion Delay). *For every arrival time $0 \leq x \leq y \leq \tau$, the congestion function f must satisfy the following condition:*

$$f(x) + x \leq f(y) + y$$

79

Such a condition will guarantee that an arriving vehicle cannot simply wait for congestion delay to reduce, and then pass a vehicle that arrived earlier. An example congestion function satisfying this property is depicted in Figure 8.

4.2 Background Literature

Relatively little research has considered time-constrained routing and scheduling problems in networks with time-varying travel times. Ahn and Shin (1991) studies the vehicle routing problem with time windows with the added constraint of time-varying congestion. The time-varying congestion is implemented by representing the internode travel time as a function of the departure time of the head node of the arc. Assuming a first-in first-out (FIFO) property for arc travel time functions (*i.e.*, a vehicle entering an arc after another vehicle will always exit later), this work develops efficient extensions of insertion, savings, and exchange-based routing heuristics for these problems. A set of feasibility check routines are developed to extend the standard heuristics to the problem with time-varying congestion.

Malandraki and Dial (1996), Hill and Benton (1992) and Malandraki and Daskin (1992) consider time-varying travel times in vehicle routing problems. These papers consider step functions for time-dependent travel time, which cannot by definition satisfy the FIFO property unless they are non-decreasing. This can be seen easily; if the travel time decreases in a future interval, then a vehicle may wait until the interval starts and then proceed to the destination and if the waiting time is less than the difference in travel times, than a violation of FIFO occurs. Another drawback of the research presented in these papers is the fact that the congestion step functions only used a few different step intervals over the course of the planning horizon.

Haghani and Jung (2005) studies a time-dependent VRP with a continuous travel speed function. The paper develops a mixed integer linear programming model for the TDDVRP to develop lower bounds. Heuristic techniques based on genetic algorithm approach are developed. Although the authors claim that the method can handle any time-dependent function, tests are performed only for piecewise linear functions with up to 30 time intervals.

Ichoua et al. (2003) studies a time-dependent VRPTW and develops heuristic solution techniques based on tabu search. The travel time function considered satisfies FIFO, but the planning period is divided into only three intervals. The results show that time-dependent model provides significant improvements in terms of effective planning for their vehicles.

This dissertation extends the existing research in this area by focusing more narrowly on drayage routing problems. Although many of the problems we consider can be formulated as time-dependent m-TSPTW problems, we develop specialized exact and heuristic techniques for problems where port congestion is the only source of cost time-dependency.

4.3 Complexity of Congested Drayage Problems

In this section, we extend complexity results for the uncongested problems UDVP and UDTP to the congested problem framework with simple proofs. Let CDVP and CDTP represent the congested drayage problems with minimizing the number of vehicles and minimizing the total travel time of all vehicles respectively.

Theorem 4.3.1. *CDVP and CDTP are NP-hard.*

Proof. In section 3.3, we prove that both the variants of UDP, namely, UDVP and UDTP, are NP-Hard. We know that problem UDVP (UDTP) is an instance of problem CDVP (CDTP), where the value of $f(x) = 0$. Hence, both CDVP and CDTP are NP-hard by restriction. □

4.4 Mixed Integer Programming formulation

We now extend the uncongested formulation developed in Section 3.4 to model a congestion delay function for vehicles arriving at the port. Since modeling general delay functions is not possible within a linear framework, we assume in this section that the delay function $f(t)$ is assumed to be a piecewise-linear function of time-of-day with a finite number of breakpoints.

Let arc set $\mathcal{A}_O$ contain all exporter-to-port and depot-to-port importer arcs. Congestion is experienced at all nodes $i \in \mathcal{N}_O$ where $\mathcal{N}_O = \{i \in \mathcal{N} \mid i = tail(a \in \mathcal{A}_O)\}$. Let c_i be the variable congestion delay time experienced by the vehicle arriving at $i \in \mathcal{N}_O$.

81

Suppose that we are given a piecewise-linear congestion delay function f defined on the interval $[0, \tau]$ with m segments. Segment j is defined by the interval $[\tau_{j-1}, \tau_j]$ where $\tau_0 \equiv 0$ and $\tau_m \equiv \tau$. The slope of the function in segment j is α_j. For convenience, define $\beta_j \equiv f(\tau_{j-1})$. Let s_i^j be the variable arrival time of a vehicle at node i if the time falls in segment j. Further, let binary variable δ_i^j be defined as:

$$
\delta_i^j = \begin{cases} 1 & \text{if the vehicle arrives at } i \text{ during segment } j \\ 0 & \text{otherwise} \end{cases}
$$

We now present the congested drayage formulation. Note that we use for clarity a number of definitional equality constraints that can be substituted out when solving. For definitions of the parameters, sets, and variables, see Section 3.4.

Formulation 2: Congested Drayage Problem Framework (CDP)

$$(\text{CDTP}) \text{ minimize} \quad \sum_{a \in \mathcal{A}} t_a x_a + \sum_{i \in \mathcal{N}_O} c_i$$

or

$$(\text{CDVP}) \text{ minimize} \quad \sum_{a \in \mathcal{O}_D} x_a \tag{32}$$

subject to:

$$\sum_{a \in \mathcal{I}_i} x_a - \sum_{a \in \mathcal{O}_i} x_a \;=\; 0 \qquad \forall \; i \in \mathcal{N} \tag{33}$$

$$\sum_{a \in \mathcal{O}_D} x_a \;\leq\; n_e + n_i \tag{34}$$

$$s_{head(a)} + t_a - M(1 - x_a) \;\leq\; s_{tail(a)} \quad \forall \; a \in \mathcal{A} \mid head(a) \notin \{D\} \cup \mathcal{N}_O \tag{35}$$

$$s'_{head(a)} + t_a - M(1 - x_a) \;\leq\; s_{tail(a)} \quad \forall \; a \in \mathcal{A} \mid head(a) \notin \{D\} \cup \mathcal{N}_O \tag{36}$$

$$t_a - M(1 - x_a) \;\leq\; s_{tail(a)} \quad \forall \; a \in \mathcal{A} \mid head(a) = D \tag{37}$$

$$a_i \leq \; s_i \; \leq \; b_i \qquad \forall \; i \in \mathcal{N} \setminus \{D\} \tag{38}$$

$$x_a \;=\; 1 \qquad \forall \; a \in \mathcal{A}_C \tag{39}$$

$$s_D \;\leq\; \tau \tag{40}$$

$$s'_i \;=\; s_i + c_i \quad \forall \; i \in \mathcal{N}_O \tag{41}$$

$$s_i \;=\; \sum_{j=1}^{m} s_i^j \quad \forall \; i \in \mathcal{N}_O \tag{42}$$

$$\tau_{j-1}\delta_i^j \leq s_i^j \;\leq\; \tau_j \delta_i^j \qquad \forall \; i \in \mathcal{N}_O \tag{43}$$

$$c_i \;=\; \sum_{j=1}^{m} \left((\beta_j - \alpha_j \tau_{j-1})\delta_i^j + \alpha_j s_i^j \right) \quad \forall \; i \in \mathcal{N}_O \tag{44}$$

$$\sum_{j=1}^{m} \delta_i^j \;=\; 1 \qquad \forall \; i \in \mathcal{N}_O \tag{45}$$

$$x_a \;\in\; \{0,1\} \quad \forall \; a \in \mathcal{A} \tag{46}$$

$$\delta_i^j \;\in\; \{0,1\} \quad \forall \; i \in \mathcal{N}_O, \; j = 1...m \tag{47}$$

$$s_i \;\geq\; 0 \qquad \forall \; i \in \mathcal{N} \tag{48}$$

$$c_i, s_i^j \;\geq\; 0 \qquad \forall \; i \in \mathcal{N}_O, \; j = 1...m \tag{49}$$

Constraints (33) through (40) repeat the uncongested formulation, with the exception of (36) which treats the departure time s'_i separately from the arrival time s_i at each congested node in $\mathcal{N}_O$. Constraint (41) determines the departure time for each congested node by adding the congestion delay c_i given by (44) to the arrival time. Constraints (42) and (43) set the interval-specific arrival time for each congested node, and (45) guarantees that only one such time is chosen.

The formulation as written requires a binary variable for each congestion function break-point for each congested node; thus, this formulation may become less useful with more complex congestion functions.

4.5 Set Covering Models for Problems CDVP and CDTP

Since the integer programming model developed in Section 4.4 is likely to be difficult to solve for large instances or when the congestion function includes many breakpoints, we again choose to implement a solution approach based on set covering models for the problems CDVP and CDTP. Since the impact of congestion delay only affects the travel times within individual routes, the set covering models developed in Section 3.5 can be applied directly in this case, where the UDVP model is used for CDVP and the UDTP model is used for CDTP. Further, we restrict attention in this case to problems where the time window for each customer request is only active at the destination of the task.

4.5.1 Using a Root Column Generation Heuristic with LSP for Solution of CDP

We again will use the root column generation heuristic to determine near-optimal solutions to the covering models. The procedures developed in Chapter 3 are used here, with the only difference arising in how to solve the column generation subproblems. In this section, we will focus on the problem CDTP; a similar extension for CDVP follows trivially.

The pricing subproblem of the column generation procedure is to identify a negative reduced-cost route to add to the current candidate list $\mathcal{R}'$. Such routes need to be time-feasible with respect to all of the problem time window constraints, and delays resulting from congestion potentially affect this feasibility.

Fortunately, it is not difficult to incorporate such delays within the LSP heuristic for finding paths. In fact, the heuristic can handle any representable congestion function: continuous or discrete, piecewise-linear, etc. The feasibility constraints developed in Section 3.6.1.1 are easily extended to incorporate the time of delay into feasibility calculations. Given the time of arrival t of the vehicle at the port, the waiting time is simply calculated from the congestion function $f(t)$ during Step 2 of the $checkExtend(\ell_{ki}, j)$ function. Similarly, the delay time is also included in the calculation of the total travel time associated with a new route generated using these techniques. We also note that it is also simple to similarly extend the enumeration technique for solving the set covering models exactly.

To illustrate the incorporation of waiting time delay into the feasibility calculations, we now formally present the LSP heuristic technique for problem CDTP. First, we note that we use the same label definition ℓ_{kj} described in section 3.7.1; importantly, note that the arrival time τ_{kj} of the path at the destination of task j will include the port delay for completing task j. Further, we will use the same network travel time definition $\bar{t}_{ij}$, but note that delay times will be added when appropriate to determine correct arrival times at different network locations.

The arc reduced costs for problem CDTP are modified to capture the delay times appropriately:

$$
\bar{c}_{ij} = \begin{cases}
\bar{t}_{ij} + d(\tau_{ki} + \bar{t}_{ij}) - \pi_j & \text{if } i \in \mathcal{V},\ j \in \mathcal{E} \\
\bar{t}_{ij} + d(\tau_{ki} + \bar{t}_{ij} - t_{Pj}) - \pi_j & \text{if } i \in \{D\} \cup \mathcal{I},\ j \in \mathcal{I} \\
\bar{t}_{ij} - \pi_j & \text{if } i \in \mathcal{E},\ j \in \mathcal{I} \\
\bar{t}_{ij} & \text{otherwise}
\end{cases}
\tag{50}
$$

Note that since delay is experienced when the vehicle reaches the port, the appropriate time of day needs to be determined differently if the task j is an import or an export task. Further, since vehicles that serve export tasks immediately followed by import tasks do not leave the port facility, they experience no additional delay.

LSP Heuristic

Initialization:

1: Initialize label ℓ_{0,v_0} corresponding to the start node v_0 representing an empty path, set all attributes of ℓ_{0,v_0} to 0

2: $\mathcal{L}^A \rightarrow \emptyset$

Iterations:

3: **for all** $j \in \mathcal{V} \setminus \{v_0\}$ **do**

4:    **if** $checkExtend(\ell_{0,v_0}, j) = \text{TRUE}$ **then**

5:      $layerExtend(\ell_{0,v_0}, j)$

6:    **end if**

7: **end for**

8: $k = 2$

9: **while** $k \leq |\mathcal{V}| - 1$ **do**

10:    **for all** $i \in \mathcal{V} \setminus \{v_0\}$ such that $\ell_{k-1,i}$ exists **do**

11:      **for all** $j \in \mathcal{V} \setminus \{v_0\}$ **do**

12:        **if** $checkExtend(\ell_{k-1,i}, j) = \text{TRUE}$ **then**

13:          $layerExtend(\ell_{k-1,i}, j)$

14:        **end if**

15:      **end for**

16:    **end for**

17:    $k = k + 1$

18: **end while**

    $checkExtend(\ell_{ki}, j)$

1: **if** $p_{ki}^j = 0$ **then**

2:    **if** $j \in \mathcal{E}$ **then**

3:      **if** $\tau_{ki} + \bar{t}_{ij} + f(\tau_{ki} + \bar{t}_{ij}) \leq b_j$ **then**

4:        **if** $\max(a_j, \tau_{ki} + \bar{t}_{ij} + f(\tau_{ki} + \bar{t}_{ij})) + \bar{t}_{jD} \leq \tau$ **then**

5:          feasIndicator $= \text{TRUE}$

6:        **end if**

7:      **end if**

8:    **else**

9:      **if** $\tau_{ki} + \bar{t}_{ij} + f(\tau_{ki} + \bar{t}_{ij} - t_{Pj}) \leq b_j$ **then**

10:        **if** $\max(a_j, \tau_{ki} + \bar{t}_{ij} + f(\tau_{ki} + \bar{t}_{ij} - t_{Pj})) + \bar{t}_{jD} \leq \tau$ **then**

11:          feasIndicator $= \text{TRUE}$

12:        **end if**

13:      **end if**

14:   **end if**

15:   **if** feasIndicator = TRUE **then**

16:     **if** $\ell_{k+1,j}$ does not exist **then**

17:       Generate blank label $\ell_{k+1,j}$

18:       return TRUE

19:     **else**

20:       **if** $\delta_{ki} + \bar{c}_{ij} \leq \delta_{k+1,j}$ **then**

21:         return TRUE

22:       **end if**

23:     **end if**

24:   **end if**

25: **end if**

26: return FALSE

$layerExtend(\ell_{ki}, j)$

1: $\mathbf{p}_{k+1,j} = \mathbf{p}_{ki}$

2: $p^{j}_{k+1,j} = 1$

3: $\delta_{k+1,j} = \delta_{ki} + \bar{c}_{ij}$

4: **if** $j \in \mathcal{E}$ **then**

5:   $\tau_{k+1,j} = max(a_j, \tau_{ki} + \bar{t}_{ij} + d(\tau_{ki} + \bar{t}_{ij}))$

6: **else**

7:   $\tau_{k+1,j} = max(a_j, \tau_{ki} + \bar{t}_{ij} + d(\tau_{ki} + \bar{t}_{ij} - t_{Pj}))$

8: **end if**

9: **if** $\delta_{k+1,j} + \bar{c}_{jD} < 0$ **then**

10:   $\mathcal{L}^A \rightarrow \mathcal{L}^A \cup \{\ell_{k+1,j}\}$

11: **end if**

With this alternative LSP pricing heuristic defined, the remainder of the root column generation solution heuristic is identical to that described in Chapter 3.

4.6   Computational Results

Solution techniques for the congested drayage problem are tested on randomly-generated problem instances representative of typical container port operations. The solution methods were implemented in the C programming language, and utilize the CPLEX Version 8.0 callable libraries for the solution of linear and binary integer programs when necessary. All tests were run on a dual-CPU 2.4 GHz Pentium with 2 GB of memory running Linux.Computation times in the tables to follow are given in seconds.

4.6.1   Data generation

Again, we assume that all container move requests face a common task completion deadline of t_d. Since all move requests face a common deadline, there are many feasible routes and therefore these problems are in some sense the most difficult that need to be solved in practice. We again generate 10 random instances on a rectangular region for each problem size, as described in the previous chapter, with travel times given by the L_1 distance function. For completeness, we repeat the problem parameters:

Rectangular service region: $x \in [0, 2]$, $y \in [0, 1]$

Depot location: $(1, \frac{1}{2})$

Port location: $(1, 0)$

Task completion deadline $= t_d = 6$ hours

Vehicle depot deadline $= \tau = 9$ hours

Note again that the dimensions of the service region are measured in hours of travel time.

We now summarize the results. All CPU times in the tables to follow are reported in seconds.

4.6.2   Impact of Congestion on Productivity

In order to analyze the impact of congestion on drayage operations, we use a simple two-segment piecewise linear delay function (see Figure 9). It is natural to assume that congestion builds toward the early part of the afternoon, and therefore we model the single peak

88

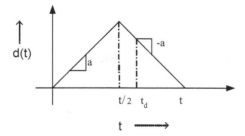

Figure 9: Congestion function for preliminary results.

at that time. The delay function is given mathematically by:

$$f(t) = \left\{ \begin{array}{ll} \alpha t & \forall t \leq \frac{\tau}{2} \\ \alpha(\tau - t) & \text{for } \frac{\tau}{2} < t \leq \tau \end{array} \right\} \tag{51}$$

where α denotes the slope of the congestion function prior to the breakpoint $\frac{\tau}{2}$ associated with the peak congestion waiting time. The slope changes to $-\alpha$ after the breakpoint. Note that this delay function satisfies a first-in, first-out condition. In the test problems, we use $\alpha = \frac{2}{9}$ hours/hour, resulting in peak delay of 1 hour.

With the generated test problem data sets, we first evaluate the effect of congestion on optimized local drayage operations at a port by solving identical instances with and without congestion delays. The results for seven 100-request problems are presented in Table 49. Note that the results in the columns labelled 'without congestion' result from solving the identical instance with $f(t) = 0$ for all t. This table clearly shows that congestion at the port is an important issue when it comes to designing optimal routes for the drayage company. Congestion delays resulted in 29.5% more travel time and required 35.4% more vehicles on an average, representing significant costs to drayage firms.

We also study the value of using an exact congestion delay function when planning. To do so, we compare the results in Table 49 to the case when the drayage company plans routes using only the average congestion waiting time at the port. Averaging only over the time that the port is open ($[0, 6]$), the average congestion delay is $\frac{7}{12}$ hours. Thus, during the planning phase, we use $f(t) = \frac{7}{12}$ for all t. Table 50 presents the results for the same seven

Table 49: Impact of Congestion Delays in 100-request Problems

Instance Number	W/o Congestion		With Congestion known	
	Travel Time	No. of Vehicles	Travel Time	No. of Vehicles
1	126.10	20	163.6	28
2	124.59	20	160.9	28
3	126.79	21	164.5	29
4	136.20	23	177.3	31
5	118.37	20	152.9	26
6	130.12	22	169.0	30
7	127.31	21	163.9	27

Table 50: Planning with Average Congestion Information

Data Set Number	Avg Congestion		Actual Congestion	
	Travel Time	No. of Vehicles	Travel Time	Deadlines Violated
1	162.8	25	162.87	7
2	160.5	26	159.48	8
3	160.5	25	161.99	11
4	175.2	28	172.63	11
5	152.5	25	152.31	3
6	168.0	27	167.26	5
7	166.5	27	164.45	6

100-request problems. The first two columns present the estimated total time required and the total number of vehicles required to serve all requests optimally; these results come from the planning phase using the average congestion information. The third column shows the actual time required to operate the planned routes, assuming the actual congestion function (51). While these actual operating times are quite similar to the predicted times and are typically better than the times presented in Table 49, we note in the fourth column that on average 7% of the customers are no longer served feasibly by the time deadline. Therefore, a company would need to buffer the average waiting time in order to use this strategy in practice.

4.7   Conclusions & Contributions

The primary contributions of this chapter include

- A review of existing literature focusing on the vehicle routing problem with time-dependent travel times, a generalization of the basic problem considered in this chapter;

- Development of decision support models for drayage routing and scheduling problems that incorporate known time-dependent waiting time for port access, including a mixed-integer programming formulation for the special case where the time-dependent delay function is piecewise-linear;

- Extension of both exact and heuristic solution approaches for large, realistic-sized drayage routing and scheduling problems with congestion. These approaches are capable of handling any time-dependent delay function that satisfies a practical FIFO property; and

- A computational study on the impact of port access delays on drayage operations, and the importance of incorporating expected delays into planning decisions.

CHAPTER V

PORT DRAYAGE WITH ACCESS SLOT CAPACITY

In this chapter, we consider a drayage routing and scheduling problem constrained by a time-slot port access control system. Under such a system, a drayage firm has limited capacity for drayage vehicle accesses to the port at different times of day. To motivate the study of this problem, we first describe the characteristics of common port access control systems. Then, we formally define the mathematical optimization problem that we will consider for the remainder of this chapter. Finally, we discuss solution approaches for the problem based on modifications of the methods developed for generalized routing problems with resource constraints presented in Chapter 2.

5.1 Introduction

In this chapter, we focus on port access congestion and the impact of access congestion management systems. Before access congestion grew as a significant problem, all ports in the U.S. allowed unscheduled access by drayage trucks and most still operate this way. In such systems, trucks may arrive to pick up and drop off containers any time within gate operating hours. Unscheduled systems clearly may be inefficient for the terminal operator, since there may be certain time periods during the day when resources are idle, and others when capacity is exceeded. Further, excessive queuing caused by unscheduled access policies has begun to worry government officials in areas with major seaports, mostly due to the environmental hazard created by diesel engine emissions from idling trucks waiting outside port gates.

One proposed remedy for the environmental and productivity problems generated by unscheduled access are drayage truck appointment systems. Recently, the state legislature in California passed the so-called "Lowenthal Bill" (Assembly Bill 2650), which limits the allowable time trucks can idle in and around port terminals to 30 minutes. Terminal operators can avoid fines associated with this restriction if they commit to the deployment of a drayage appointment system meeting certain specifications; of course, one would also hope that such a system would reduce truck queuing.

In response, many California terminals have now implemented an appointment system. Recent trade magazine articles have reported benefits of such systems, including reduced wait time for motor carriers, reduced operational costs for terminals from improved gate efficiency, accelerated throughput, and better equipment utilization. For drayage firms, these benefits hopefully lead to an increase in daily "turns" per vehicle (revenue-generating moves) and enhanced profitability. However, given the additional constraints such systems place on operations, careful planning is required to attain the maximum benefit.

In this chapter, we develop an optimization-based framework for determining high quality vehicle routes and schedules for a drayage firm operating given time-slot access capacity restrictions. We note that this framework could be used by a drayage firm attempting to select a set of drayage access appointments. We denote this problem as ACDSP. The same framework can also be used to optimize operations given a preselected set of appointments, which is denoted as ACDRP. After describing the characteristics of common port access control systems, we formally define the mathematical optimization problem that we will consider for the remainder of this paper. Next, we present a heuristic solution methodology from the problem. Finally, we discuss the computational performance of the approach, and use it to develop insight into the impact of time-slot access capacity restrictions on the efficiency of drayage operations.

5.1.1 Port Access Control Systems and Efficiency

In an effort to reduce diesel emissions from idling trucks waiting outside port gates, the California legislature enacted as law Assembly Bill (AB) 2650, commonly known as the "Lowenthal Bill" after key proponent Assemblyman Alan Lowenthal. AB 2650 restricts the allowable time trucks can idle in queue in and around port terminals to 30 minutes. Each individual violation results in a $250 fine. In effect for 2003, an additional provision of AB

2650 encourages California ports to implement access appointment systems as a means of compliance (Lowenthal, 2002).

Specifications for the appointment system in the law indicate that it must satisfy the following requirements:

1. Provide appointments on a first-come, first-served basis.

2. Provide appointments that last at least 60 minutes and are continuously staggered throughout the day.

3. Not discriminate against any motor carrier that conducts transactions at the marine terminal in scheduling appointments.

4. Not interfere with a double transaction once inside the gate.

5. Not turn away or fine a motor carrier if that motor carrier misses an appointment.

Since the passing of AB 2650, many large container port terminals in California have implemented truck access appointment systems. In general, they have similar features. Drayage firms can access the appointment systems via web interfaces, and appointment requests are processed in first-come, first-serve order. Appointments are made for hour-long time windows, and can be made up to two weeks in advance of the access day. The only information that the drayage firm must provide when scheduling an appointment is an identification code for the trucking firm. Therefore, appointments can be made in advance and the drayage firms can later decide which move requests (either inbound or outbound) to serve with each appointment.

Motivated by the proliferation of appointment-based access control systems to mitigate congestion, we study the problem of how truck drayage firms can maximize productivity when serving a port with such a system. The primary problem on which we focus is now described. Consider a single drayage firm operating a truck fleet based at a depot location, serving a single port location and a set of surrounding customer locations, and a set of container move requests to and from the port to be served on a single day. Further, suppose that the port has specified a set of access time slots, representing a discrete set of time intervals during the day, and that for each time slot, the drayage firm is limited by an upper bound on the number of vehicle accesses hereafter known as the *slot capacity*. Then, the goal of the drayage firm is to develop a set of vehicle routes and schedules that satisfy all container move requests with minimum transportation cost given the time slot capacities.

The problem framework outlined above has two primary applications for drayage companies operating under access control. In the first application, suppose that the drayage company has made a set of appointments for its vehicles in advance. This appointment set essentially defines a set of time slot capacities, and the drayage firm must optimize its operations given this set. In the second application, suppose that a firm is planning operations for a future period. Suppose that the firm has complete information regarding the remaining appointment times available for selection at the port for that future period, as well as full information regarding the container moves it must serve. After determining a minimum cost set of routes and schedules, the firm can submit appointment requests for the appropriate time slots.

5.2   Problem Definition

In this section, we formalize the problem to be studied in this chapter. Recall the problem framework UDP developed in Chapter 3. A drayage firm must serve a set $\mathcal{C}$ of container move requests in a single operating period (*e.g.*, day), in which the request set $\mathcal{C}^E$ are export container requests and the set $\mathcal{C}^I$ are import requests ($\mathcal{C}^E \cup \mathcal{C}^I = \mathcal{C}$). Each request minimally contains an origin pickup location and a destination dropoff location, one of which is the port location P. The drayage firm operates a fleet of vehicles based at a single depot location D, and develops routes and schedules for its vehicles given the set of move requests. No vehicle may depart D before time 0 and all vehicles must return to D by the time τ_D representing the end of the operating period. While problems may also be constrained with additional time windows at the origin and destination of each task, we assume for simplicity in this chapter that no such constraints are present.

The UDP problem framework can be extended to the access-controlled drayage problem

(ACDP) framework by adding time slot capacity information. Let $\mathcal{T}$ be a discrete set of non-overlapping time slots τ_j. Each $\tau_j \in \mathcal{T}$ can be described as an interval $[a_j, b_j]$ where $0 \leq a_j < b_j$. Suppose without loss of generality that the time slots are ordered such that $a_1 < a_2 < \ldots < a_{|\mathcal{T}|}$. Then, the non-overlapping property implies that $b_j \leq a_{j+1}$ for all $j = 1, \ldots, |\mathcal{T}| - 1$. Finally, each slot τ_j has a corresponding access capacity of κ_j that constrains the number of vehicle arrivals to the port within that slot in a feasible solution. Note that a vehicle might arrive to the port without a container (to pick up an import container), or with an export container; each of these types is considered an arrival and requires an available unit of time slot capacity. On the other hand, vehicles that arrive with an export container and leave after picking up an import container only require a single unit of slot capacity. The ACDP problem is now to determine a set of feasible routes and schedules, minimizing the fleet size required to serve the customer move requests. We note in this case that the vehicle departure time from each customer request is now *scheduled* such that no port time slot access capacity is exceeded. We assume that a vehicle can wait idle at a customer location either before or after picking up (or dropping off) a container.

5.3 Set Covering Models for ACDP

Since slot capacities and time constraints for drayage problems will likely lead to relatively small numbers of feasible vehicle routes, we develop set covering integer programming formulations for problem ACDP. The solution procedure involves two phases: in Phase I, we formulate and solve an integer programming model to determine which tasks we can feasibly cover, and in Phase II, we formulate the main set covering model which finds the minimum cost set of routes. While Phase I can be solved to optimality for large problems without resorting to specialized techniques, for Phase II we again determine near-optimal solutions using a root column generation heuristic.

5.3.1 Phase I: Determining Feasible Customer Requests

Given a set of customer requests $\mathcal{C}$ and a set of capacitated time slots $\mathcal{T}$ for port access, it may not be possible to serve all customer requests even with an arbitrarily large vehicle fleet. To determine a subset $\mathcal{C}'$ of feasible requests that can be covered, we formulate and solve an integer programming model that minimizes a sum of penalty costs accrued for uncovered requests.

Consider a set $\mathcal{R}^0$ of vehicle routes and schedules, where each $r \in \mathcal{R}^0$ represents a feasible sequence of customer requests to be served by a single vehicle departing and returning to the depot, satisfying customer and depot time window constraints as well as access slot capacities. Let α_{ir} be a $\{0,1\}$ parameter equal to one if request i is served by route r, and let β_{jr} represent the number of port accesses in time slot τ_j required by route r. Let parameter p_i be the penalty cost parameter for not serving request i; in this research, we set p_i to be the revenue for request i which we assume is proportional to its travel time. The decision variables y_i indicate which tasks in $\mathcal{C}$ will be feasibly served and x_r indicate which routes in $\mathcal{R}^0$ are chosen for the optimal subset.

The Phase I integer optimization problem can now be written as:

$$\text{minimize} \quad \sum_{i \in \mathcal{C}} p_i y_i$$

subject to:

$$\sum_{r \in \mathcal{R}^0} \alpha_{ir} x_r + y_i \geq 1 \qquad \forall \ i \in \mathcal{C}$$

$$\sum_{r \in \mathcal{R}^0} \beta_{jr} x_r \leq \kappa_j \qquad \forall \ \tau_j \in \mathcal{T}$$

$$x_r \in \{0,1\} \quad \forall \ r \in \mathcal{R}^0$$

We note that this optimization problem only minimizes the penalty costs for uncovered requests, and does not consider fleet size costs. Since this is the case, if we let $\mathcal{R}^0$ be the set of all so-called single-access routes, we will achieve the same objective function value if we instead considered all potential single-vehicle routes. A single access route is one where the vehicle visits the port exactly once (and therefore consumes one unit of some time slot access capacity) during its tour. The set $\mathcal{R}^0$ includes all single customer request routes ($D \rightarrow E \rightarrow P \rightarrow D$ or $D \rightarrow P \rightarrow I \rightarrow D$) for each feasible time slot, and all exporter-importer paired routes ($D \rightarrow E \rightarrow P \rightarrow I \rightarrow D$) for each feasible time slot.

5.3.2 Phase II: Optimal Route Selection

Let $\mathcal{C}'$ ($n = |\mathcal{C}'|$) be the set of feasible move requests determined after solving the Phase I model. Further, let $\mathcal{R}$ be the exponentially-large set of all feasible single-vehicle routes and schedules serving subsets of $\mathcal{C}'$; we note that certain vehicle route sequences may be repeated multiple times to reflect different possibilities for time slot access. The following standard set covering model selects a subset of $\mathcal{R}$ that ensures that each customer request in $\mathcal{C}'$ is served, while minimizing the total number of required vehicles:

$$\text{minimize} \quad \sum_{r \in \mathcal{R}} x_r$$

subject to:

$$\sum_{r \in \mathcal{R}} \alpha_{ir} x_r \geq 1 \quad \forall\; i \in \mathcal{C}' \tag{52}$$

$$\sum_{r \in \mathcal{R}} \beta_{jr} x_r \leq \kappa_j \quad \forall\; \tau_j \in \mathcal{T} \tag{53}$$

$$x_r \in \{0, 1\} \quad \forall\; r \in \mathcal{R} \tag{54}$$

5.3.2.1 Comparison of Objectives

The formulation for the ACDP presented above minimizes the size of the vehicle fleet required to serve all of the requests in $\mathcal{C}'$. Since access slots are capacity-constrained, it is also natural to consider the alternative objective of minimizing the number of required slot accesses to serve all customer requests. We note that this alternative objective function can be written as $\sum_{r \in \mathcal{R}} (\sum_{\tau_j \in \mathcal{T}} \beta_{jr}) x_r$; when this alternative objective is used, denote the problem ACDP-A. The following theorem holds:

Theorem 5.3.1. *The set of all optimal solutions for ACDP need not contain the optimal solution for ACDP-A, and vice versa.*

Proof. We prove this using a counter-example. Consider an instance of the the drayage problem with three customer locations, where each customer location submits one export move request and one import move request. Let the travel time between the depot and any of the customer locations or the port be one time unit. The travel time between the port and any of the customer locations is two time units. Suppose the only time constraint on the system is that all vehicles must return to the depot by a deadline of 10 time units; there are no slots or slot capacity constraints.

Given this setting, a vehicle operating a single request route will require 4 time units, while a vehicle serving an export-import pair at the same customer location will require 6 units. Clearly, the optimal solution to problem ACDP-A is to use three vehicles, each serving such an exporter-importer pair and therefore requiring three vehicle accesses to the port.

On the other hand, the optimal solution to problem ACDP in this case requires only 2 vehicles. The first vehicle serves the export-import pair at customer location 1, and the export request at location 2 thus requiring 10 time units and two port accesses. The second vehicle similarly serves the export-import pair at customer location 3, and the import request at location 2 also requiring 10 time units and two port accesses. Since this solution requires 4 port accesses, it is not optimal to problem ACDP-A. □

We believe that the problem of determining the minimum fleet requirements needed to serve a set of requests is more relevant than the problem of determining the minimum number of required port accesses, so we will proceed with a study of problem ACDP.

5.3.3   Using a Root Column Generation Heuristic with LSP for Solution of ACDP

Enumeration of the set of feasible routes $\mathcal{R}^0$ required by the Phase I integer programming model is a simple task with polynomial complexity. This is not the case when enumerating $\mathcal{R}$ for the Phase II model, since the number of feasible routes may grow exponentially in the number of customer move requests. Since $\mathcal{R}$ will contain a very large number of routes for instances of practical size, we again develop a root column generation solution heuristic for the problem.

In the root column generation heuristic, we first determine a near-optimal solution to linear programming relaxation of the Phase II integer programming model using column generation. As explained in detail in Chapter 2, at each iteration of this approach, a restricted version of the linear relaxation of the Phase II model is solved using a subset $\mathcal{R}' \subset \mathcal{R}$ of the complete set of feasible vehicle routes and schedules. Using optimal dual variable information from the solution of this restricted problem, a pricing subproblem is then solved to identify routes and schedules in $\mathcal{R} \setminus \mathcal{R}'$ with negative reduced cost that therefore would improve the solution of the linear relaxation. If such routes are identified, they are added to $\mathcal{R}'$ and the process is repeated until no such routes are found. At this point, the Phase II integer program is solved with the final route subset $\mathcal{R}'$.

It can also be shown rather easily that a good initial set of routes and schedules is $\mathcal{R}' = \mathcal{R}^0$:

Observation 1. *A problem instance of ACDP with feasible customer request set $\mathcal{C}'$ has a feasible solution if and only if the Phase II formulation has a feasible solution using the subset of routes and schedules $\mathcal{R}^0$.*

We again will utilize a heuristic algorithm to determine near-optimal solutions to the pricing subproblem within the column generation procedure. To do so, we extend the LSP heuristic developed in Chapter 2. The goal of the pricing subproblem is to identify columns with negative reduced cost to add to the current column subset considered by the master linear program. We have already demonstrated in Chapter 2 that the LSP heuristic is an efficient and effective procedure to generate near-optimal solutions for ESPPRC subproblems associated with column generation.

5.3.3.1 ACDP Column Generation Subproblem

We now detail the pricing subproblem to be solved to identify routes and schedules with negative reduced costs each iteration. Let π_i and σ_j represent the dual variables associated with constraints (52) and (53) after solving the linear relaxation of the Phase II integer program using column subset $\mathcal{R}'$. Then, the reduced cost $\bar{c}_r$ of a route $r \in \mathcal{R}$ is given by

$$\bar{c}_r = 1 - \sum_{i \in \mathcal{C}} \alpha_{ir}\pi_i - \sum_{\tau_j \in \mathcal{T}} \beta_{jr}\sigma_j \tag{55}$$

The column generation subproblem for the drayage problem is then:

$$\min_{r \in \mathcal{R} \setminus \mathcal{R}'} \bar{c}_r. \tag{56}$$

If the solution to this problem has a negative objective function value, then we have identified a negative reduced cost route and schedule to add. In the ACDP context, the above problem is a time-dependent elementary shortest-path problem with time-window constraints. The time dependency results from the fact that the dual variable for a port access changes over time, and therefore the cost of a route depends on the time(s) that it is scheduled to access the port. As noted in Chapter 2, Dror (1994) provides a detailed discussion regarding the complexity of this NP-hard shortest-path problem.

5.3.3.2 LSP Pricing Heuristic for ACDP

To apply the LSP heuristic to the ACDP pricing subproblem, we extend the network representation to use space-time nodes for each customer request at each feasible port access time slot. Therefore, the heuristic maintains at most $(n_e + n_i)|T|$ labels in each of maximum possible $(n_e + n_i)$ layers; this contrasts to the original approach for UDP problems that maintains a single label for each request in each layer, for a maximum of $(n_e + n_i)^2$ labels.

To formally specify the heuristic approach, we begin by defining the network $G = (\mathcal{V}, \mathcal{A})$ over which the method will determine paths. Each node $v_i^j \in \mathcal{V}$ represents a single container move request i to be served using a specific port access time slot $\tau_j \in \mathcal{T}$, and a node for the container depot: $\mathcal{V} = \{D\} \cup (\mathcal{E} \times \mathcal{T}) \cup (\mathcal{I} \times \mathcal{T})$. Arcs connecting all locations v_i^p and v_j^q in $\mathcal{V}$ such that $q \geq p$ are created to form $\mathcal{A}$. To solve the ACDP problem, we need to only consider a single resource type, time. The definition for time consumption along each arc (v_i^p, v_j^q), given by $\bar{t}_{v_i^p, v_j^q}$, is similar to the definition in section 3.7.1 for UDP problems. For the sake of completeness, we repeat it here:

$$
\bar{t}_{v_i^p, v_j^q} = \begin{cases}
t_{Dj} + t_{jP} & \text{if } v_i^p = D, \ v_j^q \in (\mathcal{E} \times \mathcal{T}) \\
t_{DP} + t_{Pj} & \text{if } v_i^p = D, \ v_j^q \in (\mathcal{I} \times \mathcal{T}) \\
t_{Pj} + t_{jP} & \text{if } v_i^p \in (\mathcal{E} \times \mathcal{T}), \ v_j^q \in (\mathcal{E} \times \mathcal{T}) \\
t_{Pj} & \text{if } v_i^p \in (\mathcal{E} \times \mathcal{T}), \ v_j^q \in (\mathcal{I} \times \mathcal{T}) \\
t_{ij} + t_{jP} & \text{if } v_i^p \in (\mathcal{I} \times \mathcal{T}), \ v_j^q \in (\mathcal{E} \times \mathcal{T}) \\
t_{iP} + t_{Pj} & \text{if } v_i^p \in (\mathcal{I} \times \mathcal{T}), \ v_j^q \in (\mathcal{I} \times \mathcal{T}) \\
t_{PD} & \text{if } v_i^p \in (\mathcal{E} \times \mathcal{T}), \ v_j^q = D \\
t_{iD} & \text{if } v_i^p \in (\mathcal{I} \times \mathcal{T}), \ v_j^q = D
\end{cases}
\tag{57}
$$

Now we can define the arc reduced costs as follows:

$$
\bar{c}_{v_i^p, v_j^q} = \begin{cases}
0 & \text{if } v_i^p \in (\mathcal{E} \times \mathcal{T}) \cup (\mathcal{I} \times \mathcal{T}), \ v_j^q = D \\
1 - \pi_j - \sigma_{\tau_q} & \text{if } v_i^p = D, \ v_j^q \in (\mathcal{E} \times \mathcal{T}) \cup (\mathcal{I} \times \mathcal{T}) \\
-\pi_j & \text{if } v_i^p \in (\mathcal{E} \times \mathcal{T}), \ v_j^q \in (\mathcal{I} \times \mathcal{T}) \quad \text{and} \quad \tau_q = \tau_p \\
-\pi_j - \sigma_{\tau_q} & \text{otherwise}
\end{cases}
\tag{58}
$$

The start node for path generation is the vehicle depot: $v_0 = D$. Again, each label ℓ_{k, v_j^q} represents the layered shortest path $P^*_{v_0, v_j^q}(k)$ serving request j as its last request, accessing the port in slot τ_q for request j, after completing $k - 1$ prior requests. One additional label $\ell_{0 v_0}$ is used for initialization. Let $\mathcal{L}$ denote the set of all labels ℓ_{k, v_j^q} generated by the heuristic, and $\mathcal{L}^A$ be the set of labels corresponding to negative reduced cost paths.

For the ACDP, we augment the contents of each label to include a feasible port access time window for the last request served by the route; this window is by definition a subset of the window defined by the time slot utilized for this request. This time window enables simple determination of feasible route extensions to future task-slot pairs. Thus, each label $\ell_{k, v_j^q} \in \mathcal{L}$ contains the following attributes:

- A path vector $\mathbf{p}_{k,v_j^q}$ of length $n_e + n_i$ with each element

$$p_{k,v_j^q}^i = \begin{cases} 1 & \text{if customer request } i \in \mathcal{E} \cup \mathcal{I} \text{ is already covered in path } P_{v_0,v_j^q}^*(k) \\ 0 & \text{otherwise} \end{cases} ;$$

- A time slot vector $\mathbf{ts}_{k,v_j^q}$ with each element

$$ts_{k,v_j^q}^{\tau_i} = \text{number of port accesses in time slot } \tau_i \in \mathcal{T} \text{ required by path } P_{v_0,v_j^q}^*(k)$$

- The feasible port access time window, denoted by $[a', b']_{k,v_j^q}$, for request j during time slot τ_q, the final request served by path $P_{v_0,v_j^q}^*(k)$.

- The cost of path $P_{v_0,v_j^q}^*(k)$, given by δ_{k,v_j^q}.

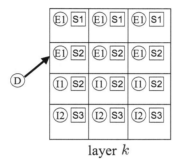

layer k

Figure 10: Portion of the label array structure for LSP pricing heuristic; each entry represents a route that serves its final request as indicated by the circle during the time slot indicated by the square.

LSP Heuristic

Initialization:

1: Initialize label ℓ_{0,v_0} corresponding to the start node v_0 representing an empty path, set all attributes of ℓ_{0,v_0} to 0

2: $\mathcal{L}^A \to \emptyset$

Iterations:

103

3: **for all** $v_j^q \in \mathcal{V} \setminus \{v_0\}$ **do**

4:     **if** $checkExtend(\ell_{0,v_0}, v_j^q) = \text{TRUE}$ **then**

5:         $layerExtend(\ell_{0,v_0}, v_j^q)$

6:     **end if**

7: **end for**

8: $k = 2$

9: **while** $k \leq |\mathcal{V}| - 1$ **do**

10:     **for all** $v_i^p \in \mathcal{V} \setminus \{v_0\}$ such that ℓ_{k-1,v_i^p} exists **do**

11:         **for all** $v_j^q \in \mathcal{V} \setminus \{v_0\}$ **do**

12:             **if** $checkExtend(\ell_{k-1,v_i^p}, v_j^q) = \text{TRUE}$ **then**

13:                 $layerExtend(\ell_{k-1,v_i^p}, v_j^q)$

14:             **end if**

15:         **end for**

16:     **end for**

17:     $k = k + 1$

18: **end while**

$checkExtend(\ell_{k,v_i^p}, v_j^q)$

1: **if** $p_{k,v_i^p}^j = 0$ **then**

2:     **if** $a_{\tau_p} \leq a_{\tau_q}$ **then**

3:         $(a', b') = calculatePortWindow(\ell_{k,v_i^p}, v_j^q))$

4:         **if** $a' \leq b'$ **then**

5:             **if** $a' \leq b_{\tau_q}$ **then**

6:                 **if** $b' \geq a_{\tau_q}$ **then**

7:                     **if** ℓ_{k+1,v_j^q} does not exist **then**

8:                         Generate blank label ℓ_{k+1,v_j^q}

9:                         return TRUE

10:                     **else**

11:                         **if** $\delta_{k+1,v_j^q}^{new} \leq \delta_{k+1,v_j^q}^{old}$ **then**

12:                             return TRUE

13:           **end if**

14:          **end if**

15:        **end if**

16:      **end if**

17:    **end if**

18: **end if**

19: **end if**

20: return FALSE

$layerExtend(\ell_{k,v_i^p}, v_j^q)$

1: $(a'_{k+1,v_j^q}, b'_{k+1,v_j^q}) = calculatePortWindow(\ell_{k,v_i^p}, v_j^q))$

2: $\mathbf{p}_{k+1,v_j^q} = \mathbf{p}_{k,v_i^p}$

3: $p_{k+1,v_j^q}^{j} = 1$

4: $\mathbf{ts}_{k+1,v_j^q} = \mathbf{ts}_{k,v_i^p}$

5: $ts_{k+1,v_j^q}^{\tau_j} = ts_{k+1,v_j^q}^{\tau_j} + 1$ unless $i \in \mathcal{E}$ and $j \in \mathcal{I}$

6: $\delta_{k+1,v_j^q} = \delta_{k,v_i^p} + \bar{c}_{v_i^p,v_j^q}$

7: **if** $\delta_{k+1,v_j^q} + \bar{c}_{v_j^q D} < 0$ **then**

8:    $\mathcal{L}^A \to \mathcal{L}^A \cup \{\ell_{k+1,v_j^q}\}$

9: **end if**

$calculatePortWindow(\ell_{k,v_i^p}, v_j^q)$

Case I : $i \in \mathcal{E}$ and $j \in \mathcal{E}$

- $a' \leftarrow \max(a'_{k,v_i^p} + 2t_{Pj}, a_{\tau_j})$

- $b' \leftarrow \min(\tau_D - t_{PD}, b_{\tau_j})$

Case II: $i \in \mathcal{I}$ and $j \in \mathcal{E}$

- $a' \leftarrow \max(a'_{k,v_i^p} + t_{Pi} + t_{ij} + t_{jP}, a_{\tau_j})$

- $b' \leftarrow \min(\tau_D - t_{PD}, b_{\tau_j})$

Case III: $i \in \mathcal{E}$ and $j \in \mathcal{I}$

- $a' \leftarrow \max(a'_{k,v_i^p}, a_{\tau_j})$

- $b' \leftarrow \min(\tau_D - t_{Pj} - t_{jD}, b_{\tau_j})$

Case IV: $i \in \mathcal{I}$ and $j \in \mathcal{I}$

- $a' \leftarrow \max(a'_{k,v_i^p} + 2t_{Pi}, a_{\tau_j})$

- $b' \leftarrow \min(\tau_D - t_{Pj} - t_{jD}, b_{\tau_j})$

return(a', b')

At the conclusion of the heuristic, we add all labels in the set $\mathcal{L}^A$ corresponding to routes with negative reduced costs to $\mathcal{R}'$. If $\mathcal{L}^A = \emptyset$, we have found no improving routes and thus the heuristic column generation is complete.

5.4 Computational Experiments on Impact of Appointment Systems

In this section, we present the results of a set of computational experiments for the access-controlled port drayage problem. First, we seek to understand the computational efficiency of the heuristic approach for problems of practical size. Further, we attempt to understand the relationship between the productivity of a drayage firm and the characteristics of the set of appointments available on a given operating day. Specifically, we will investigate the productivity impact of the total number of available accesses, the distribution of the number of accesses across time slots, the time slot duration and distribution of customers across the region.

To this end, we test and report results for randomly-generated problem instances representative of typical container port operations. The solution approach is implemented in the C programming language, and utilizes the CPLEX Version 8.1 callable libraries for the solution of linear and binary integer programs when necessary. All tests were run on a dual-CPU 2.4 GHz Pentium with 2 GB of memory running Linux. Computation times in the tables to follow are given in seconds.

In Chapter 2, we have already shown that the root column generation heuristic can generate high quality solutions for VRPTW problems. Further, Chapter 3 shows that the method is particularly effective on large unconstrained drayage problems. With this motivation, we assume that the quality of the heuristic solutions is reasonable. As will be seen, the solution approach allows solution of instances of ACDP with up to 100 container drayage requests within 20-25 minutes of CPU time. The method, therefore, appears to be practical for typical problem sizes faced in the industry.

5.4.1 Data Generation

In this chapter, we consider two separate data sets. The first data set uses the customer locations generated for the results in Section 3.8, and considers the same operating time window at the port. We also develop a second data set in this study, where customers are on average located further from the depot and port but the operating hours of the facilities are longer. We explain the generation of this data set below.

Assume that the fleet depot of the drayage company operates from 6 AM until 8 PM; *i.e.*, vehicles can start from the depot as early as 6 AM and can return to the depot as late as 8 PM. We also assume that the port and all customer warehouses, both import and export, allow pickups or deliveries from 8 AM to 6 PM. To again consider difficult-to-solve drayage problems, we assume that all container move requests should be completed only within this operating day; no additional time windows constrain the requests. Since there are therefore many feasible vehicle routes and schedules, these problems are among the most difficult that need to be solved in practice.

We generate 10 different problem instances, each with 100 customer requests, from each of 50 exporter and 50 importer locations. To generate an instance, the location coordinates of each exporter and importer warehouse are randomly generated using a two-dimensional uniform distribution specified over a service region. The location of the port and the depot are fixed at two points inside the rectangular region for all problem data sets. Travel times are specified by an L_1 distance function.

The following parameters are used for all generated problem instances:

- Square service region: $x \in [0, 4]$ hours, $y \in [0, 4]$ hours

- Depot location D: $(2, 3)$

107

- Port location P: $(2, 1)$

Let 8 AM represent time $t = 0$ for our problems. To model the port access control system, the available access time interval at the port, $[0, 10]$, is partitioned into $|T|$ equal-duration slots. In this study, we consider experiments with three different slot durations SD: 30 minutes, 60 minutes and 120 minutes. The corresponding values for $|T|$ are 20, 10 and 5 respectively. Further, we compare three different assumptions regarding the distribution of available port access capacity across the slots. Figure 11 depicts the general form of the three distributions. The V-shaped function, denoted *Morning/Afternoon Heavy*, represents the case where the drayage firm has selected more time slots in the morning and afternoon, and fewer around the midday. It is generated as follows. First, the total access slot capacity SC is divided by the number of slots $|T|$ to yield the average per period capacity AC (note that this may be non-integer). Next, a slope s representing the per period change in capacity is determined by assuming that the average capacity is $\frac{AC}{2}$ at time zero and $\frac{3AC}{2}$ at time $\frac{|T|}{2}$. Using these points, and slope s before time $\frac{|T|}{2}$ and $-s$ after time $\frac{|T|}{2}$, capacities are generated for each time slot. Slot capacities are then rounded to integer values such that the total capacity of all slots sums to SC. A similar procedure is used to generate the upside-down V-shaped function denoted *Midday Heavy*; in this case the capacity of $\frac{3AC}{2}$ is at time zero, and $\frac{AC}{2}$ at time $\frac{|T|}{2}$. For the first data set, the port is only open for six hours. We use a similar methodology to build slot capacities for this case, however we only consider a slot duration of one hour.

Figure 11: Graphical depiction of the three port access time slot capacity functions; example with total capacity $TC = 52$ and slot duration $SD = 120$ minutes.

Total	No. of	CPU Time			
Capacity(SC)	Columns	Phase I	Heuristic	B & B	Total
46	45965.8	38.1	89.15	132.67	259.94
48	47050.7	47.4	89.38	298.44	435.2
50	50522.2	61.9	110.7	545.83	718.39
52	55188.5	49	130.4	482.28	661.7
54	55988.6	16.8	124.8	868.25	1009.8
56	58545.8	0	108.2	1240.6	1348.9
58	57548.9	0	102.1	500.42	602.47
60	60151.3	0	104.8	783.11	887.95

Table 51: Performance of LSP Heuristic using Uniform capacity distribution and $SD = 30$ minutes

5.4.2 Solution Computation Time Performance

First, we summarize the solution computation time performance. For brevity, we only present computation time results for problems drawn using the second customer data set; results for the first data set are similar. Tables 51 to 59 summarize the performance of the solution approach for ACDP instances for each of the three slot capacity distributions and for each of the slot durations. Each row provides average computation time statistics over the 10 sample instances described earlier. The first column reports the total slot access capacity SC available. The second column reports the average number of columns generated by the LSP subproblem heuristic when solving the root linear program of Phase II. The third, fourth and fifth columns report the average CPU execution time required to solve the Phase I integer program to optimality, the heuristic column generation using LSP for the Phase II root linear program, and then the Phase II branch-and-bound. The sixth column provides the total execution time required to generate a complete solution.

Consider Tables 51, 52 and 53 which summarizes the performance of the solution approach for the uniform, mid-day heavy and morning/afternoon heavy distributions; when the slot duration is $SD = 30$ minutes. It is easy to observe that as the total available capacity SC increases, the number of candidate routes generated increases, and the instances require more time to solve. But, even for the most difficult instances, complete solutions were generated using no more than 21 minutes of CPU time on average. Thus, the methodology appears to be quite effective for finding solutions to practical, real-world

Total	No. of	CPU Time			
Capacity(SC)	Columns	Phase I	Heuristic	B & B	Total
46	49766.6	116.7	102.2	131.3	350.23
48	54200.4	143	129.8	511	783.82
50	59273.5	105	156.5	216.4	477.92
52	64674.8	0	158.4	627	785.39
54	66067.7	0	143.4	295	438.46
56	63329.6	0	139.5	466.2	605.66
58	64034.5	0	121.6	927	1048.6
60	64543.9	0	112.6	160.6	273.16

Table 52: Performance of LSP Heuristic using Mid-day Heavy capacity distribution and $SD = 30$ minutes

Total	No. of	CPU Time			
Capacity(TC)	Columns	Phase I	Heuristic	B & B	Total
46	42965.9	12.7	69.05	329.09	410.86
48	44616.6	28.5	79.91	273.02	381.41
50	46627.9	32.2	84.14	298.79	415.12
52	47245.5	16.4	86.22	376.95	479.55
54	49781.4	19.7	102.3	574.76	696.7
56	53532.9	12.6	126.6	861.05	1000.3
58	53510.1	2.16	96.79	1148.9	1247.9
60	54885.3	0	95.41	1041.6	1137

Table 53: Solution Computation Time: Morning/Afternoon Heavy Capacity Distribution and $SD = 30$ minutes

instances within reasonable computation times. Also, each of the Phase I problems were solved within 2-3 minutes, justifying the use of a mixed integer programming model solved exactly via branch-and-bound to determine the tasks to be served.

Comparing the results in the tables, it should be noted that the problem instances with

Total	No. of	CPU Time			
Capacity(SC)	Columns	Phase I	Heuristic	B & B	Total
46	24876.7	4.8	28.5	76.35	109.7
48	25315.7	4.9	31.3	78.32	114.6
50	27431	3.5	37.8	141.5	182.8
52	30607.6	1.3	42.1	227.1	270.5
54	30770.1	0.2	41.4	535.7	577.4
56	32938.6	0	38.8	580.1	619
58	33414.9	0	37.8	440.3	478.1
60	32480.9	0	34	393.4	427.4

Table 54: Performance of LSP Heuristic using uniform capacity distribution and $SD = 60$ minutes

Total	No. of	CPU Time			
Capacity(SC)	Columns	Phase I	Heuristic	B & B	Total
46	26273.5	12.4	34.8	44.81	92.07
48	28986.3	26.1	43.6	169.8	239.5
50	32903.6	6.27	58	123.4	187.7
52	32240.5	0.22	47.6	330.3	378.2
54	36034.1	0	48.3	417	465.3
56	35522.2	0	47.9	125.1	173
58	35038.6	0	39.9	154	193.9
60	34515.7	0	36.1	97.17	133.2

Table 55: Performance of LSP Heuristic using Mid-Day Heavy capacity distribution and $SD = 60$ minutes

Total	No. of	CPU Time			
Capacity(SC)	Columns	Phase I	Heuristic	B & B	Total
46	23013.2	1.7	22.5	41.53	65.69
48	23708.1	1.5	26	86.93	114.5
50	24445.9	3.8	27.5	52.3	83.64
52	26020.7	2.9	32.1	323.6	358.5
54	25655.3	2.6	31.1	142.7	176.4
56	27596.2	0.2	40.8	681.5	722.4
58	29468.9	0	32.2	706	738.1
60	29905.3	0	30.5	371.7	402.2

Table 56: Performance of LSP Heuristic using Morning/Afternoon Heavy capacity distribution and $SD = 60$ minutes

$SD = 30$ require the most computation time. This can be attributed to the fact that a single candidate route in a problem instance with $SD = 120$ minutes may result in up to 4 different candidate routes when $SD = 30$ minutes. Thus, it is reasonable to believe that problem instances will generally become more computationally-intensive as the duration of the access time slots is reduced.

5.4.3 Impact of Appointment Characteristics

We now attempt to understand the impact of appointment-based access control on the productivity of drayage firms. To do so, we use our solution approach to determine near-optimal drayage routing and scheduling plans for identical instances with varying appointment characteristics, namely total access capacity (SC), slot duration (SD), and capacity distribution. We also study the impact of the customer location characteristics by comparing the results from the two different data sets. The primary productivity performance

111

Total	No. of	CPU Time			
Capacity(SC)	Columns	Phase I	Heuristic	B & B	Total
46	14927.9	0.7	20.2	29.98	50.84
48	15735.5	2.2	15.8	22.71	40.74
50	18267.8	0.1	21.5	33.66	55.21
52	21542.8	0	22.6	287.2	309.8
54	21566.7	0	21.8	177.1	198.8
56	21735	0	20.5	117.7	138.2
58	21164.1	0	19.2	160.7	179.9
60	20893.6	0	17.7	95.57	113.3

Table 57: Performance of LSP Heuristic using Uniform capacity distribution and $SD = 120$ minutes

Total	No. of	CPU Time			
Capacity(SC)	Columns	Phase I	Heuristic	B & B	Total
46	15526.5	4.12	13.9	13.67	31.66
48	17415.6	3.09	20.3	35.6	58.95
50	21155.6	0	28.8	62.31	91.11
52	23176.7	0	26.4	292	318.4
54	23205.5	0	24.3	203.5	227.8
56	22841.4	0	25.1	136.1	161.3
58	22475.7	0	21.3	57.8	79.1
60	22283.4	0	20.4	87.57	107.9

Table 58: Performance of LSP Heuristic using Mid-day Heavy capacity distribution and $SD = 120$ minutes

Total	No. of	CPU Time			
Capacity(SC)	Columns	Phase I	Heuristic	B & B	Total
46	14249.8	0.7	11.2	19.43	31.31
48	15236.5	0.5	14	11.21	25.72
50	15896.8	0.4	15	27.25	42.68
52	16272.5	0.3	15.3	34.23	49.83
54	18364	0.1	18.5	79.94	98.51
56	18655.5	0	18.9	101.5	120.3
58	19229.2	0	16	198.9	214.9
60	19139.2	0	17.4	270.4	287.8

Table 59: Performance of LSP Heuristic using Morning/Afternoon Heavy capacity distribution and $SD = 120$ minutes

metrics will be the total number of tasks that can be feasibly served, and the total number of vehicles required to serve the feasible tasks.

Figures 12 to 19 summarize these results. In each figure, the x-axis represents the total number of port accesses available over all slots (SC).

5.4.3.1 Impact of varying distribution of customers

First, we examine the impact of different customer location distributions. One can think of two essential classes of ACDP instances, depending on the type of binding constraint causing infeasible routes. In the first class, tight capacity constraints for each time slot result in fewer customers per vehicle. In these instances, the slot capacity constraint will be binding, and any increase in slot capacity will result in significant increase in drayage firm productivity. The first data set demonstrates an example of such a class class of instances. In the second class, tight time windows for each customer request can result in fewer customers per vehicle, even if there is enough capacity available at each time slot at the port. In such problem instances, the time windows will be the binding constraint, and increasing the available port capacity may not result in significant increase in drayage firm productivity. The second data set represents this second class of instances.

Figure 12 compares the productivity impacts of varying the slot capacities under the two cases. It is easily observable that the change in the number of vehicles required to serve the set of feasible tasks is dramatic for the first data set, and is almost insignificant for the second data set. Table 60 substantiates this fact. It can be seen that as the total available slot capacity increases, the percentage increase in the total number of routes generated is much higher in the case of the first data set, showing that the slot capacity constraint is more binding in this case.

5.4.3.2 Impact of varying total access capacity

We now study further the relationship between total access capacity and productivity. Each of the figures 13 to 19 depicts how increasing the total available access capacity affects the productivity of the drayage firm. Importantly, note that it is often the case that small increases in the total capacity SC (and therefore small changes in the capacity of each

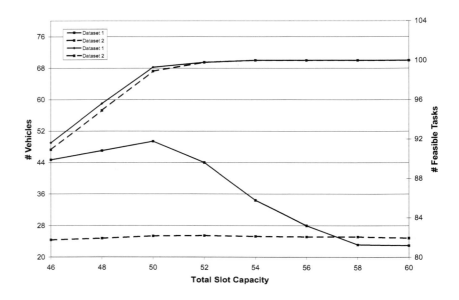

Figure 12: Productivity Impacts Depending on Customer Location Distribution: Slot Duration 60 minutes; Mid-day Heavy Capacity Distribution

Table 60: Differences in the number of feasible routes generated for two different data sets for ACDP problems

Total Capacity	Routes generated by LSP heuristic	
	Data Set 1	Data Set 2
46	10663.7	26273.5
48	11342.6	28986.3
50	11723	32903.6
52	13001.4	32240.5
54	15622	36034.1
56	19483.2	35522.2
58	22221	35038.6
60	22867.1	34515.7

time slot) can dramatically affect the number of feasible tasks when capacity is restricted. Each extra unit of capacity allows at most two additional tasks to be served, and it is clear from the figures that the actual number of additional tasks served often corresponds to this bound. Further, the number of vehicles required to perform the feasible drayage requests also changes significantly when capacity is added for smaller values of SC, but it is interesting to note that while every extra unit of capacity often allows two additional tasks to be covered, the resultant average increase in the number of required vehicles is always less than one (and usually substantially less); therefore, the additional capacity often allows the firm to reallocate its vehicles to serve the requests, and does not necessarily require an additional vehicle.

Such results indicate therefore that it is imperative that drayage firms make good port slot selections in order to maximize productivity; those who do not may suffer substantial customer service penalties by not serving requests. It is also interesting to note that when SC increases beyond the point where all customers can be served feasibly, in most cases the average number of vehicles required to serve the requests begins to decrease. If the problems were solved to optimality, and there were no rounding effects of determining individual access slot capacities, then such decreases should always exist until an unconstrained minimum number of vehicles is achieved. In these results, this decrease in the average number of vehicles required is relative small in most cases and does not exceed 0.5 vehicles per unit of capacity increase.

5.4.3.3   Impact of varying distribution of total capacity across slots

Figures 13 to 16 depict the variation in the number of feasible tasks performed, number of vehicles required and the number of port accesses used for each access capacity distribution considered, for slot duration values $SD = 30, 60$ and 120 minutes respectively. In the upper figure in each pair, the y-axis gives the number of feasible tasks (out of 100) that can be completed. Using only these feasible tasks, the lower figure shows the number of port accesses and vehicles required in the solution determined using our methodology, where we note that the number of accesses used is always no less than the number of required vehicles.From these figures, it is clear that the Mid-Day Heavy capacity distribution leads to higher drayage firm productivity than the Morning/Afternoon Heavy distribution for this set of problem instances. On an average, the latter distribution required approximately 10

115

% more total capacity to serve all requests. Further, even when there was sufficient total capacity (*i.e.*, $SC = 60$) to find a feasible solution for most instances, the Morning/Afternoon Heavy distribution required an increased vehicle fleet size to serve all requests.

An interesting observation is the fact that for smaller values of SC, *i.e.*, $SC = 46$ and 48, the Mid-day Heavy distribution requires the most vehicles. While this is primarily due to the fact that under this distribution more tasks are feasible, it is also true in when $SD = 120$ and the Mid-day and Continuous distributions both allow the same number of feasible tasks. For larger values of SC, the Mid-day Heavy distribution requires the fewest vehicles indicating that this distribution gives the drayage company the most routing flexibility.

While this paper only investigates a few different options for the distributions of access capacity, it seems clear that the drayage firm needs to be quite careful when booking access capacity within specific slots in order to maximize the customer requests that it can serve and minimize the fleet costs it will incur.

5.4.3.4 *Impact of varying length of slot duration*

Lastly, Figures 17 to 19 depict the variation in drayage firm productivity for different values of the slot duration: $SD = 30, 60$ and 120 minutes respectively. Like in the previous set of figures, the y-axis in the upper figure in each pair gives the number of feasible tasks (out of 100) that can be completed. Using only these feasible tasks, the lower figure shows the number of port accesses and vehicles required in the solution determined using our methodology.Again, it is important to note that slot duration does substantially affect the number of tasks that can be served by the drayage company, and the fleet costs of serving a set of tasks. As expected, since the wider slot duration value provides more flexibility for the drayage company, firm productivity is maximized for wider slots. Notably, the relative difference between operations under different slot durations is minimal when the distribution of slots follows the Mid-day Heavy pattern. Further, when slots are distributed using the "worst" distribution (Morning/Afternoon Heavy), the relative gain of widening the slots from 30 minutes to 120 minutes appears more than linear. Alternately, when slots are distributed using the Continuous distribution, the same relative gains appear to be less than linear. In any case, it is clear that access slot length is an important factor to consider when designing a capacitated port access system, and that ports should perform careful studies of productivity impacts before selecting an appropriate slot length.

5.5   Conclusions and Contributions

Optimization of port drayage operations becomes complex when port access is restricted by available time slot capacity. Importantly, this research shows that productivity of large drayage firms serving many daily requests can be significantly impacted by relatively minor changes in the characteristics of the allowable port accesses. Therefore, ports need to carefully consider such productivity impacts when designing a port access system. Further, drayage companies operating under such a system should seriously consider using a decision support approach such as the one described in this paper to aid in the selection of access bookings, and the optimization of operations given a selection of bookings.

The primary contributions of this research include

- Formulations of integer programming optimization models for determining optimal routes and schedules for drayage vehicles given port access time slot capacities;

- Development of a fast heuristic based on column generation that generate near-optimal solutions for port drayage routing and scheduling problems with port access time slot capacities, which should be useful in practice for both selection of port access appointments and optimization given a set of preselected appointments; and

- A computational study that indicates substantial productivity impacts of small variations in port access time slot capacities on drayage firms, measured by the number of feasible customer requests that can be served and the fleet size required to serve the set of feasible customer requests.

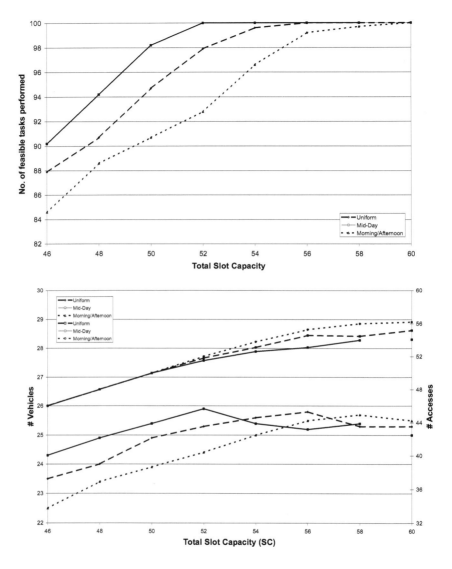

Figure 13: Productivity Impact of Total Capacity Distribution: Slot Duration 30 minutes; Data Set 2

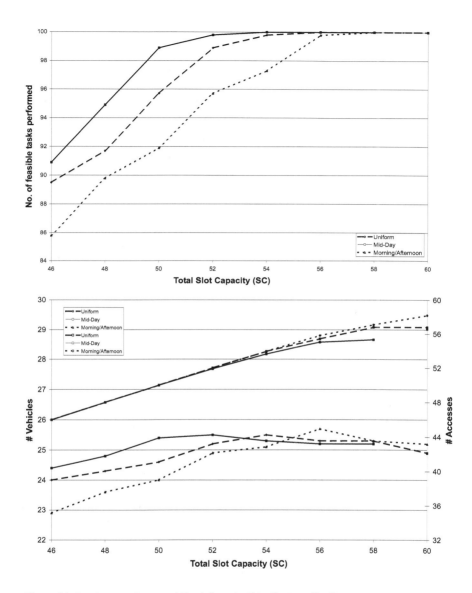

Figure 14: Productivity Impact of Total Capacity Distribution: Slot Duration 60 minutes; Data Set 2

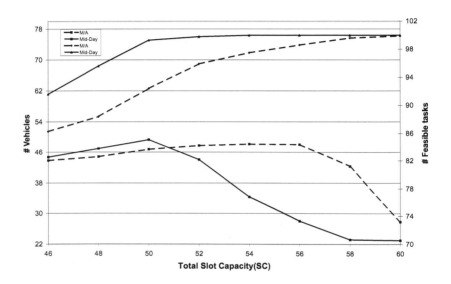

Figure 15: Productivity Impact of Total Capacity Distribution: Slot Duration 60 minutes; Data Set 1

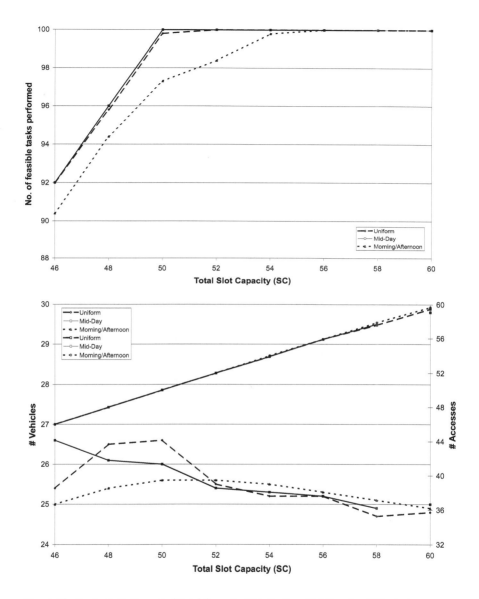

Figure 16: Productivity Impact of Total Capacity Distribution: Slot Duration 120 minutes; Data Set 2

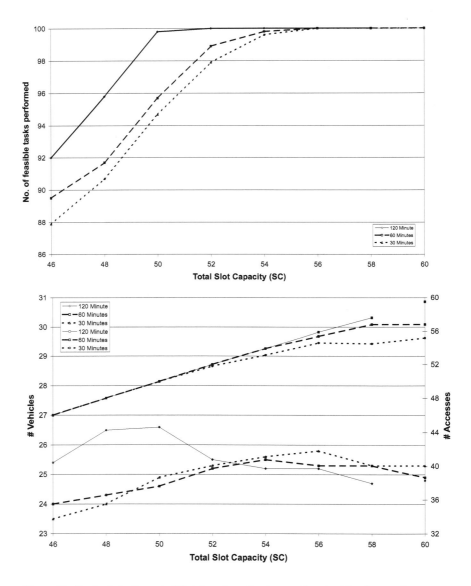

Figure 17: Productivity Impact of Slot Duration: Continuous Capacity Distribution; Data Set 2

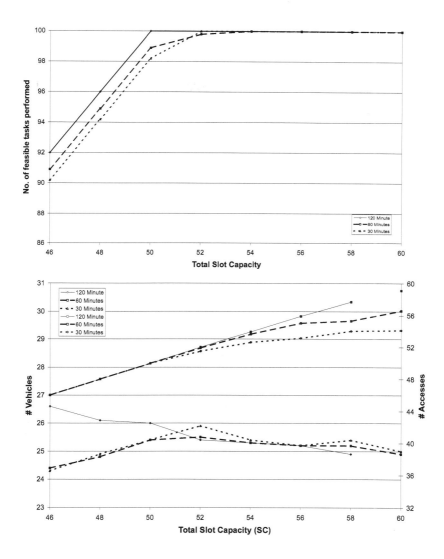

Figure 18: Productivity Impact of Slot Duration: Mid-Day Heavy Capacity Distribution; Data Set 2

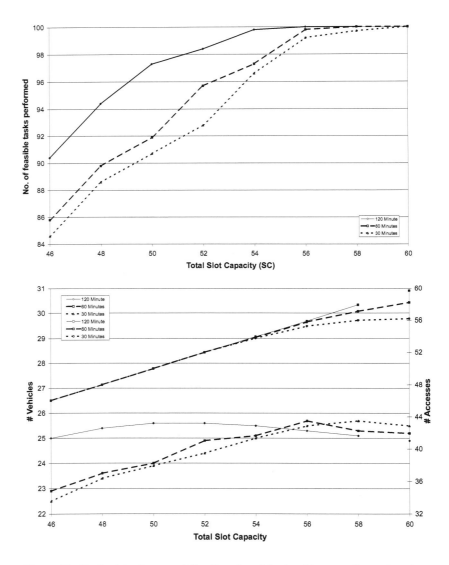

Figure 19: Productivity Impact of Slot Duration: Morning-Afternoon Heavy Capacity Distribution; Data Set 2

Bibliography

B.H. Ahn and J.Y. Shin. Vehicle-routing with time windows and time-varying congestion. *Transportation Research Part B*, 42:393–400, 1991.

J. Bramel and D. Simchi-Levi. On the effectiveness of set covering formulations for the vehicle routing problem with time windows. *Operations Research*, 45(2):295–301, 1997.

M. Gendreau D. Feillet, P. Dejax and C. Gueguen. An exact algorithm for the elementary shortest path problem with resource constraints: Application to some vehicle routing problems. *Networks*, 44(3):216–229, 2004.

G.B. Dantzig and P. Wolfe. Decomposition principle for linear programs. *Operations Research*, 8:101–111, 1960.

M. Desrochers. An algorithm for the shortest path problem with resource constraints. Technical report, GERAD, 1988. G-88-27.

M. Desrochers, J. Desrosiers, and M. Solomon. A new optimization algorithm for the vehicle routing problem with time windows. *Operations Research*, 40(2):342–354, 1992.

M. Dror. Note on complexity of shortest path models for column generation in the vrptw. *Operations Research*, 42:977–978, 1994.

Y. Dumas, J. Desrosiers, and F. Soumis. The pickup and delivery problem with time windows. *European Journal of Operational Research*, 54:7–22, 1991.

M. Gronalt, R.F. Hartl, and M. Reimann. New savings based algorithms for time constrained pickup and delivery of full truckloads. *European Journal of Operational Research*, 151:520–535, 2003.

A. Haghani and S. Jung. A dynamic vehicle routing problem with time-dependent travel times. *Computers & Operations Research*, 32:2959–2986, 2005.

A.J. Herberger. Expanding opportunities in coastwise shipping: A multi-modal, integrated coastal transportation system for the 21st century. Technical report, Testimony before the House Transportation and Infrastructure Committee, Washington, D.C., May 2001. http://www.house.gov/transportation/cgmt/05-23-01/herberger.html(last accessed May 2006).

A.V. Hill and W.C. Benton. Modeling intra-city time-dependent travel speeds for vehicle scheduling problems. *Journal of Operational Research Society*, 43:343–351, 1992.

J. Holguin-Veras. On the attitudinal characteristics of motor carriers toward container availability systems. *International Journal of Services Technology and Management*, 1 (2/3):140–155, 2000.

J. Holguin-Veras and C.M. Walton. State of the practice of information technology at marine container ports. *Transportation Research Record*, 1522:87–93, 1996.

S. Ichoua, M. Gendreau, and J.Y. Potvin. Vehicle dispatching with time-dependent travel times. *European Journal of Operational Research*, 144:379–396, 2003.

S. Irnich and G. Desaulniers. *Shortest Path Problems with Resource Constraints*, chapter 2, pages 33–65. GERAD 25th Anniversary Series. Springer, 2005.

A. Lowenthal. H&s 40720 marine terminal operation; truck idling. Technical report, California State Assembly, State of California, September 2002. http://www.arb.ca.gov/bluebook/bb03/HS/40720.htm(last accessed May 2006).

J.M. MacDonald. Port security/credentialing. Technical report, Testimony before the House Transportation and Infrastructure Committee, Washington, D.C., February 2002. http://www.house.gov/transportation/cgmt/02-13-02/macdonald.html(last accessed May 2006).

C. Malandraki and M.S. Daskin. Time-dependent vehicle routing problems: formulations, properties and heuristic algorithms. *Transportation Science*, 26(3):185–200, 1992.

C. Malandraki and R.B. Dial. A restricted dynamic programming heuristic algorithm for the time dependent traveling salesman problem. *European Journal of Operational Research*, 90:45–55, 1996.

S. Mitrovic-Minic. Pickup and delivery problems with time windows: A survey. Technical report, Simon Fraser University, SFU CMPT TR1998-12, May 1998. ftp://fas.sfu.ca/pub/cs/techreports/1998.

T.E. Rajasimhan. A moving experience. Technical report, August 2002. http://www.blonnet.com/ew/2002/08/21/stories/2002082100220300.htm(last accessed May 2006).

A.C. Regan and T.F. Golob. Trucking industry perceptions of congestion problems and potential solutions in maritime intermodal operations in california. *Transportation Research: Part A*, 34:587–605, 2000.

L.M. Rousseau, M.Gendreau, G.Pesant, and F.Focacci. Solving vrptws with constraint programming based column generation. *Annals of Operations Research*, 130(1-4):199–216, 2004.

M.W.P. Savelsbergh and M. Sol. The general pickup and delivery problem. *Transportation Science*, 29:17–29, 1995.

W.G. Schubert. The security of our seaports. Technical report, Testimony before the Senate Judiciary Committee, Subcommittee On Technology, Terrorism And Government Information, Washington, D.C., February 2002. http://www.tsa.dot.gov/public/display?theme=47&content=383(last accessed May 2006).

M.M. Solomon. Algorithms for the vehicle routing and scheduling problems with time window constraints. *Operations Research*, 35(2):254–265, 1987.

M.M. Solomon. Benchmarking problems. Personal Webpage, March 2005. http://web.cba.neu.edu/ msolomon/problems.htm.

X. Wang and A.C. Regan. Local truckload pickup and delivery with hard time window constraints. *Transportation Research: Part B*, 36:97–112, 2002.

VITA

Rajeev Namboothiri was born on July 7, 1979 in Thiruvananthapuram, Kerala, India. He received his Bachelor of Technology in Mechanical Engineering from Indian Institute of Technology, Madras, India in Spring 2001. He received his Master of Science in Industrial Engineering from Georgia Tech in Spring 2003. Rajeev's research interests are mainly in the areas of Transportation and Logistics and Supply Chain Management

36456263R00084

Made in the USA
San Bernardino, CA
22 May 2019